THE NZ SERIES

WEATHER AND CLIMATE
NEW ZEALAND

Sandra Carrod

with graphics by Karsten Schneider

Oratia

FRONT COVER GRAPHIC Karsten Schneider
Photographs courtesy of Sandra Carrod except:
contents Keith Edkins; p. 14 Isabella McFadden;
p. 18 Murray Watson; p. 26 National Science Foundation;
p. 28 NASA; p. 31 A. Zaki *NZ Herald*; p. 34 NASA;
p. 41 NASA; pp. 45 lower, 46, 49 NIWA; p. 48 NASA;
p. 50 Keith Edkins; p. 67 Chris Tarpey *NZ Herald*;
p. 82 Tim Hawkins; p. 89 bottom Shaun Gilbertson;
pp. 93, 95, 96 Max Kafka. The author acknowledges
NIWA and MetService for various data resources
from which she has drawn information.

Published by Oratia Books, Oratia Media Ltd,
783 West Coast Road, Oratia, Auckland 0604, New Zealand
(www.oratia.co.nz)

Copyright © 2022 Sandra Carrod
Copyright © 2022 Karsten Schneider — graphic illustrations
Copyright © 2022 Oratia Books (published work)

The copyright holders assert their moral rights in the work.

This book is copyright. Except for the purposes of fair reviewing,
no part of this publication may be reproduced or transmitted
in any form or by any means, whether electronic, digital or
mechanical, including photocopying, recording, any digital or
computerised format, or any information storage and retrieval
system, including by any means via the Internet, without
permission in writing from the publisher. Infringers of copyright
render themselves liable to prosecution.

ISBN 978-1-99-004226-3

Managing editor: Carolyn Lagahetau
Designer: Sarah Elworthy

Printed in China

CONTENTS

What's the difference between climate and weather? 4
Aotearoa New Zealand's place in the world 6
The atmosphere and what powers Earth's weather 9
Cycles and constants 13
Global atmospheric circulation 15
Earth's spheres 18
The Goldilocks planet 21
Weather watchers 23
Earth's climate 25
Climate modelling 30
Mending the planet 34
The Sun and space weather 38
Oceans and climate 40
Changing sea levels 45
Tropical cyclones 48
El Niño Southern Oscillation 52
Sea breezes and land breezes 57
Clouds 59
Thunderstorms 61
Tornadoes 66
How weather systems form 68
How to read a weather map 72
Wind 78
Rain 83
Snow 87
Frost 91
Fog and mist 93
Glaciers 95
The climate crisis 97
Index 100

WHAT'S THE DIFFERENCE BETWEEN CLIMATE AND WEATHER?

Climate is the usual weather an area or region has at different seasons of the year. It is the average weather — indicators like temperature, wind strength and direction, precipitation and humidity measured over decades.

Weather is the mix of events that happen in our atmosphere and can change from one minute to the next, often making it difficult to predict. Some years may be unusually hot or cold, wet or dry, but climate, the most commonly experienced weather patterns of a place, normally changes much more slowly.

Different parts of Aotearoa have very distinct climates. Distance from the Equator (*latitude*), height above sea level (*elevation*) and distance from the sea all influence the climate of a region. The North Island, closer to the Equator, is generally warmer than the South Island, yet the highest temperatures in New Zealand are often recorded in the South Island, in areas like Central Otago that are farthest from the sea. Climate is also influenced by *topography*, the physical features of the land's surface. High country can deflect airflows, both funnelling wind and giving protection from it. Mountains have a big impact on the climate of a place. The mountain ranges that run the length of much of the country bring more rain to western areas than to regions east of the Main Divide. The contrast in rainfall between western and eastern climates is much greater than the difference between the north and south.

Climate determines how we live, from our outdoor activities and the clothes we wear, to how much energy we use in our homes and the kinds of food we grow. It affects where and when the rain falls and how much; it impacts our water supplies and even the air we breathe.

MICROCLIMATES A microclimate exists in a small, localised area. They can be slightly warmer, drier or more sheltered from the wind than their surroundings. They can also be colder, wetter, windier or more frost-prone. The size of a microclimate can vary from a few square metres in a garden, to areas in a city centre. Microclimates can occur naturally or be formed by human activities. Cities often have urban microclimates, tending to have less sun and more rain and cloud, and to be warmer by both day and night than surrounding country.

AOTEAROA NEW ZEALAND'S PLACE IN THE WORLD

Aotearoa New Zealand lies in the temperate zone, midway between the warm tropics and the South Pole. It is a long, narrow country, stretching around 1600 kilometres (km) from the Far North to the Deep South and almost 450 km at its widest point. It spans almost 13 degrees of latitude, from 34°S at North Cape to 47°S at Stewart Island/Rakiura, so the climate ranges from subtropical in Northland to cool temperate down south. Surrounded by sea — the Tasman Sea to the west, the South Pacific Ocean to the north and east, and the Southern Ocean to the south — it has a mostly maritime climate, which makes our weather extremely changeable.

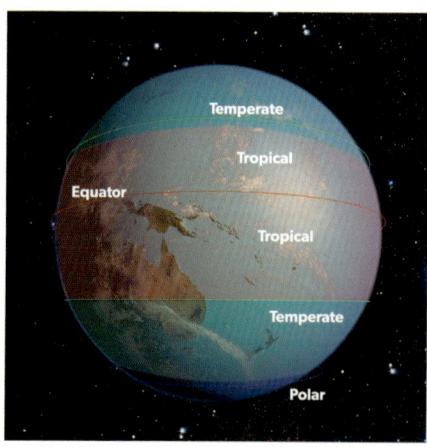

The difference in temperature between the Equator and the two poles powers the world's weather. Our weather is driven by our position in the southern hemisphere, surrounded by sea, in the temperate zone midway between the Equator and the South Pole.

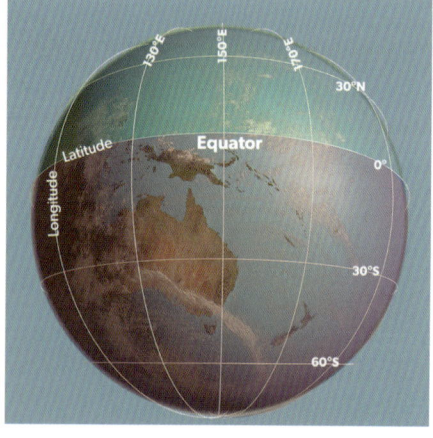

Lines of latitude run parallel to the Equator (latitude zero degrees), both north and south of the Equator, which splits the globe into two hemispheres. Lines of longitude run from pole to pole, cutting across the Equator.

AOTEAROA NEW ZEALAND'S PLACE IN THE WORLD

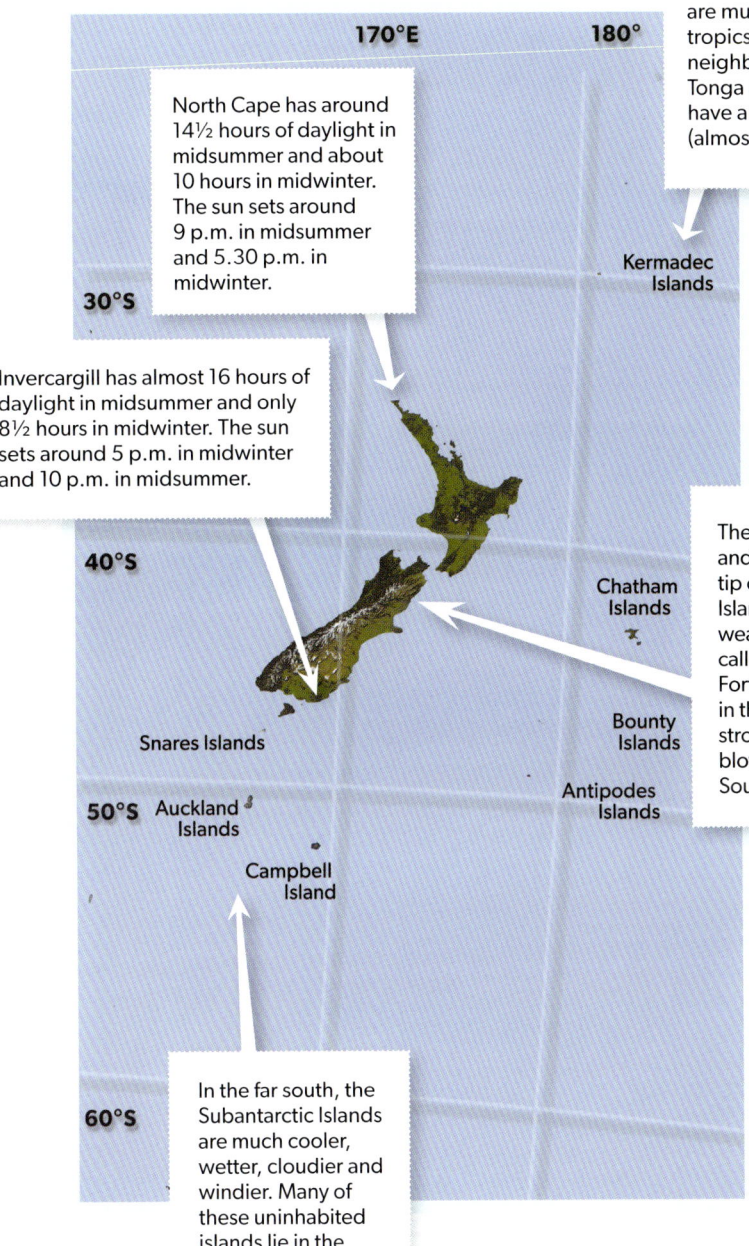

The Kermadec Islands, at 30° south of the Equator, are much closer to the tropics and our nearest neighbours to the north, Tonga and Fiji. They have a warm subtropical (almost tropical) climate.

North Cape has around 14½ hours of daylight in midsummer and about 10 hours in midwinter. The sun sets around 9 p.m. in midsummer and 5.30 p.m. in midwinter.

Invercargill has almost 16 hours of daylight in midsummer and only 8½ hours in midwinter. The sun sets around 5 p.m. in midwinter and 10 p.m. in midsummer.

The South Island and the southern tip of the North Island lie in the wild weather latitudes called the 'Roaring Forties' — right in the path of the strong winds that blow around the Southern Ocean.

In the far south, the Subantarctic Islands are much cooler, wetter, cloudier and windier. Many of these uninhabited islands lie in the stormy 'Furious Fifties'.

THE MARITIME CLIMATE The maritime climate is relatively mild and very changeable, with cooler summers and warmer winters compared to regions surrounded by land. The Chatham Islands, 800 km east of mainland New Zealand, are a good example. Rēkohu, the Moriori name for the islands, means 'misty sun'.

Auckland, a narrow isthmus (strip of land) is almost an island and equally renowned for its changeable weather, as well as its humid climate. Humidity is usually higher in the North Island because of the warm, humid air masses coming in over the ocean from the subtropics.

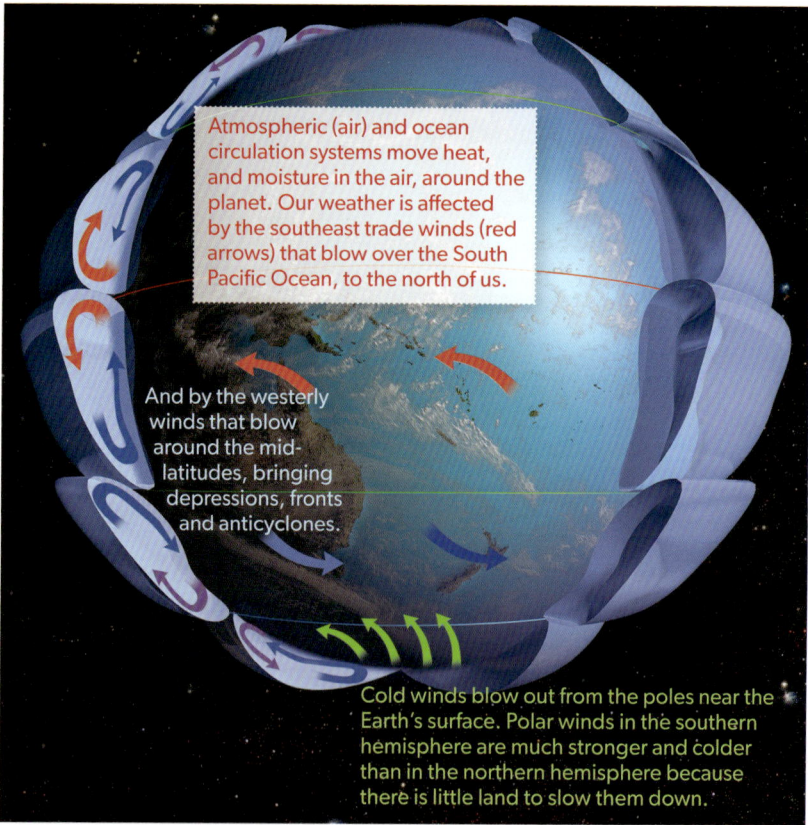

Atmospheric (air) and ocean circulation systems move heat, and moisture in the air, around the planet. Our weather is affected by the southeast trade winds (red arrows) that blow over the South Pacific Ocean, to the north of us.

And by the westerly winds that blow around the mid-latitudes, bringing depressions, fronts and anticyclones.

Cold winds blow out from the poles near the Earth's surface. Polar winds in the southern hemisphere are much stronger and colder than in the northern hemisphere because there is little land to slow them down.

Winds are named for the direction they blow *from*, so a westerly wind comes from the west and blows eastwards. Meteorologists measure wind strength at a height of ten metres above the ground because friction makes wind speed drop off towards the surface. The direction a wind blows *from* is a good indicator of the kind of weather it will bring.

THE ATMOSPHERE AND WHAT POWERS EARTH'S WEATHER

Earth's atmosphere is made up of layers of gases surrounding the planet. If you imagine Earth as a giant apple, the atmosphere would be thinner than the apple's skin, in relative terms. Our weather happens mostly in the troposphere, the layer closest to Earth's surface. The temperature, gases and air pressure of each layer are different. The air in the stratosphere is very dry. Commercial jet aircraft often fly at the lower level of this layer which is above most of the clouds and bad weather. We depend on the atmosphere to shield our planet from harmful ultraviolet and X-ray radiation and to insulate us from extreme temperature changes.

The tropopause is the boundary between the troposphere and the stratosphere. The differences in surface temperature and air pressure around the globe mean that the tropopause is higher over the Equator (around 15–18 km) than above the poles (7–10 km). For the same reason, the tropopause also varies with the seasons and is lower in winter and higher in summer. Over Aotearoa it is about 11 km high and can be 2–4 km higher in summer.

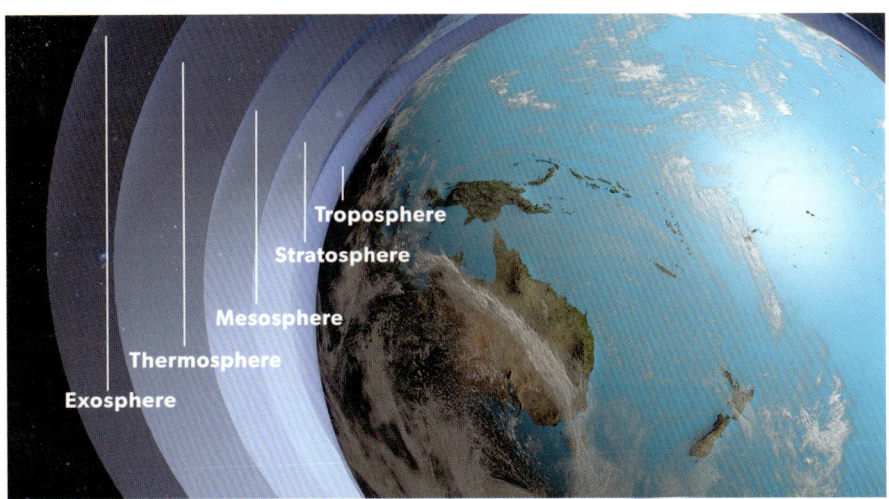

TEMPERATURE GRADIENTS

A temperature gradient is a change in temperature over distance. The amount of heat from the Sun varies, so the temperature gradient in the troposphere varies with the time of day and with place. The different layers of the atmosphere have very different temperature gradients.

> **INVERSION** Normally, the higher you go, the colder the air gets — in the troposphere at least. An inversion is the opposite (or inverse) of what you might expect: an increase of temperature with altitude.

Temperature inversion in the stratosphere puts a layer of warm air above our troposphere, acting like a cap on the weather. This prevents air rising into the stratosphere and separates the two layers. Occasionally, towering cumulonimbus clouds briefly push up into the stratosphere.

The world's weather is powered by the Sun, which heats the Earth unevenly, so that the Equator receives more sunlight than the poles. Differences in temperature and air pressure drive the movement of air around the planet. Constantly shifting warm and cold air masses redistribute heat and moisture, bringing changes in the weather.

The Sun's rays fall most directly on the Equator and have to travel further through Earth's atmosphere to reach the poles. They are stronger and more concentrated at the Equator and more widely spread in polar regions.

Water molecules move through the system in an endless cycle, changing from gas (water vapour) to liquid (water) and solid (ice) as the temperature and pressure change. Most of the water vapour present in the atmosphere has evaporated from the world's oceans.

The ocean circulation system works in tandem with the atmospheric system, absorbing heat and carbon dioxide (CO_2), and carrying nutrients from the depths of the sea to the surface waters. Living things, including humans, are a natural part of the carbon cycle, taking up carbon and releasing it to the atmosphere, by eating and breathing, just as, by drinking, sweating and urinating, we are all part of the water cycle.

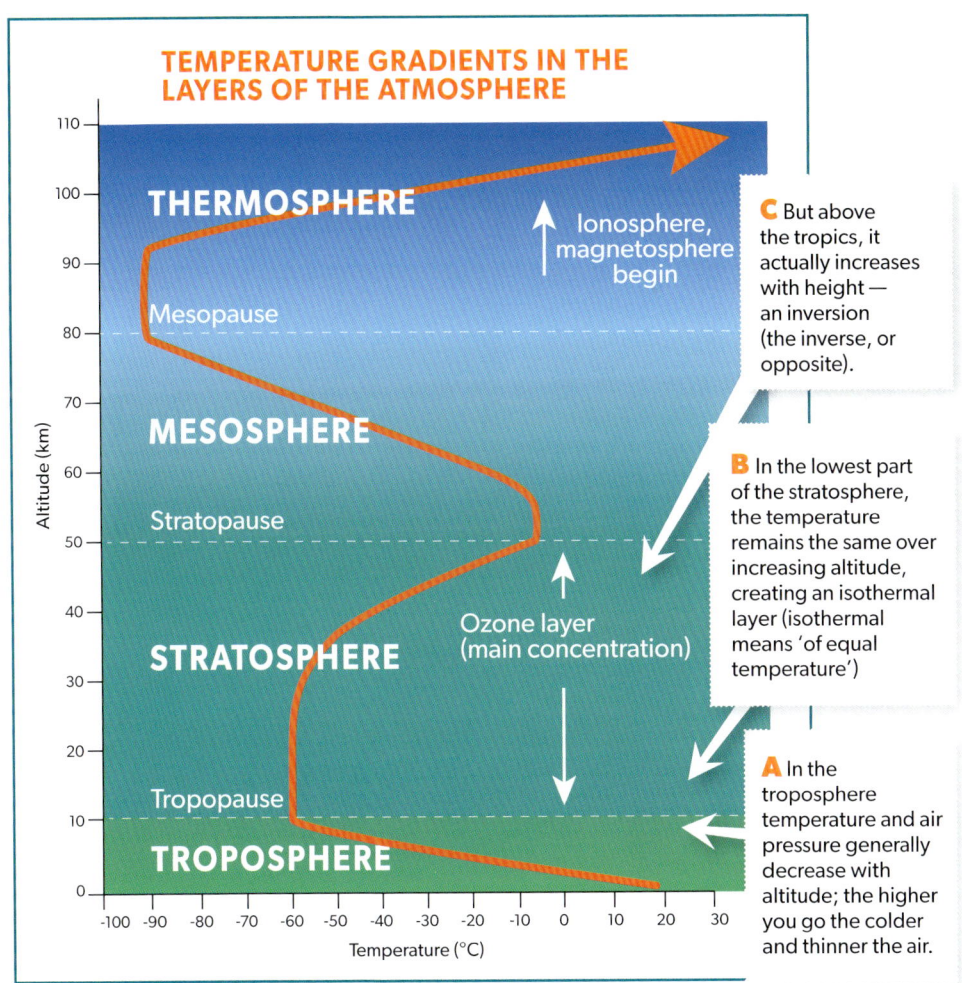

WHY IS IT COLD IN THE MOUNTAINS?

In the troposphere the air gets thinner, or less dense, the higher above the surface you go. It expands with increasing altitude, the air molecules spread further apart and so the air temperature decreases — about 6.5 degrees Celsius for every 1000 metres of height above sea level. This is called the environmental lapse rate.

The force of gravity pulls the air molecules in the Earth's atmosphere towards our planet's surface. The weight of all the air pressing down means that air pressure is greatest at sea level, where the air molecules are packed close together.

Air near Earth's surface is warmer because of the Sun's heat radiating back from the ground.

CYCLES AND CONSTANTS

EARTH'S ROTATION

The Earth spins from west to east on its axis, turning once every 24 hours, giving us night and day. The poles are the points where Earth's axis of rotation meets its surface. The axis of rotation is tilted, which heats the hemispheres differently as Earth orbits the Sun, creating the four seasons of the year.

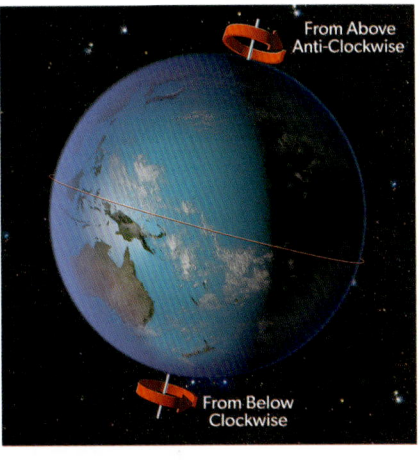

At the Equator seasons are defined more by rainfall than by amount of sunlight. In Aotearoa seasonal differences are more marked in the South Island, which is closer to the South Pole and its alternating seasons of light and darkness.

CONVECTION CURRENTS

Convection currents transfer heat from one place to another and can occur in air, water, or even in molten rock deep below Earth's surface. They drive Earth's ocean and atmospheric circulation systems, as well as much smaller, local changes in the weather. The world's major winds — the trade winds, the westerlies and the polar winds — are giant convection currents. At the other end of the scale, a convection current is hot air rising above pots

cooking on the stove and a draught of cold air moving in under a door to take its place.

THE CORIOLIS EFFECT

As the Earth turns on its axis the world's winds are pushed off course by its spin, much as a ball thrown from a spinning roundabout does not travel in a straight line but swerves from its course. At New Zealand's latitude the prevailing winds blow as westerlies, instead of blowing north towards the Equator (near the surface) and south towards the pole (at altitude) as they would without Earth's spin. The Coriolis effect causes the atmosphere's circulation system to split into three different cells of air circulating at different latitudes in each hemisphere.

> **POLAR DRIFT** Earth's current angle of tilt relative to its orbit round the sun is 23.4 degrees and decreasing. The angle naturally oscillates (wobbles) between 22.1 and 24.5 degrees in a 41,000-year cycle. Changes in how Earth's mass is spread over the planet can cause the axis, and therefore the position of the poles, to shift. This 'polar drift' occurs naturally because of changes in ocean currents and movements deep below the Earth's crust. Scientists have discovered that the rate of axis shift has sped up in the last 25 years. They believe human activities are responsible for this change and the melting of glaciers and removal of groundwater from beneath the surface have altered the distribution of water in land and sea around the globe. Once groundwater has been extracted and used for drinking or irrigation it mostly ends up in the ocean, rather than beneath land, and that changes how weight is distributed around Earth's surface.

GLOBAL ATMOSPHERIC CIRCULATION

THE EARTH'S SYSTEM

Earth's climate system is a complex web of interactions that function together as a whole, on an intricate variety of timescales, to keep our climate in balance. If any aspect of the global system gets thrown off kilter, it impacts the smooth functioning of the whole. Rocks and the surface of the planet play their part by absorbing solar energy, radiating heat and storing carbon. Land and sea reflect and hold heat differently. The changes that humans make to the environment often have unforeseen repercussions for the delicate balance. Processes deep inside the Earth, as well as at every depth of the ocean and level of the atmosphere, all play their part in regulating our environment and our climate.

THE HADLEY, FERREL AND POLAR CELLS

Three separate convection cells operate together, circulating through all the levels of the atmosphere, moving warm air away from the Equator and cool air towards it.

Jet streams are narrow bands of high-speed wind that circulate in meandering 'rivers of air' high in the atmosphere. They blow from west to east but they can wander both northwards or southwards for a time and they can abruptly stop and start, rise and fall slightly in altitude, and, like rivers, branch into separate flows. Disturbances in the usual movement of the jet streams can cause unusual changes in the weather as the normal boundaries shift between the cold polar air and warm subtropical air.

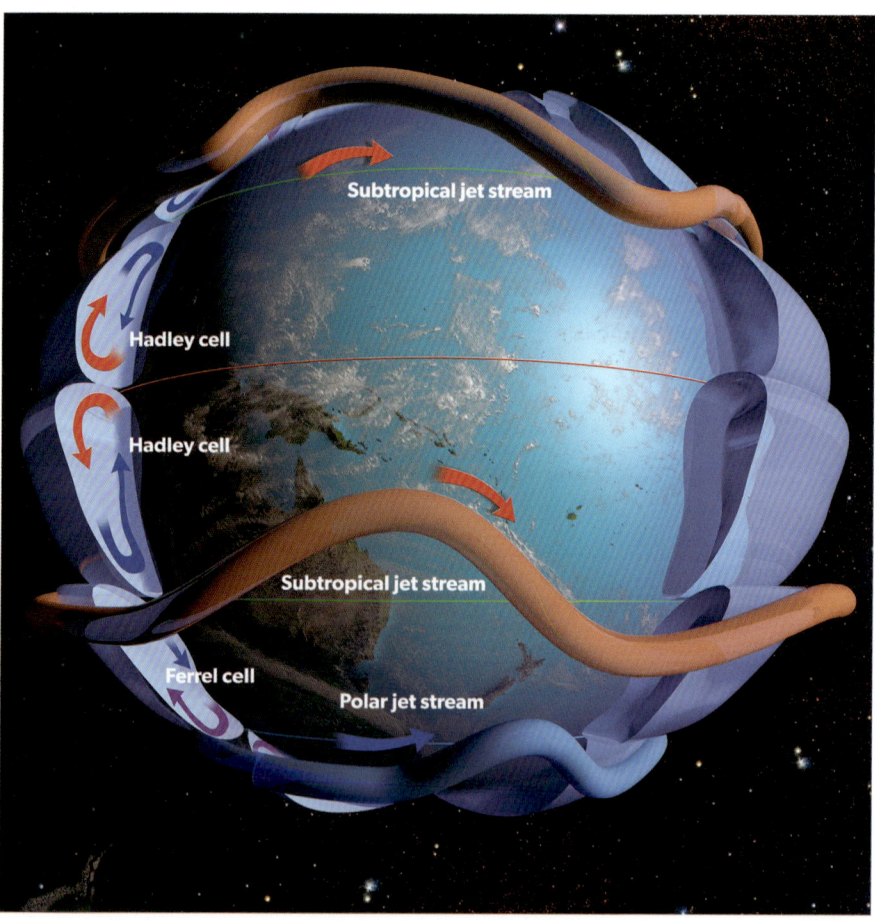

GLOBAL ATMOSPHERIC CIRCULATION

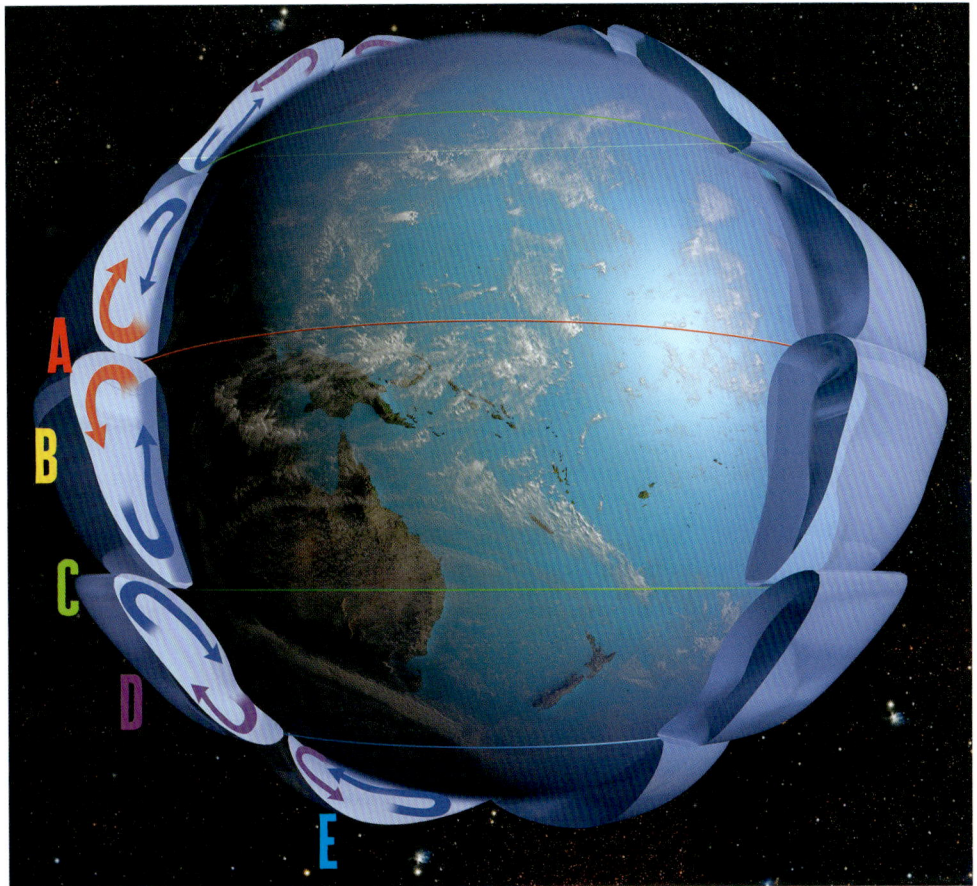

A Air above the Equator warms, expands and rises in the troposphere, just as a hot air balloon rises because the air inside it is lighter and less dense than the surrounding air.

B As the air rises high into the troposphere it cools until at about 10–15 km above Earth's surface it starts to sink back down, flowing out towards the poles.

C As it nears the surface, at about 30° South in our hemisphere, it flows back towards the Equator, warming as it does, and the whole cycle starts again.

D In the Ferrel cells between 30° and 60° south and north of the Equator, air flows in the opposite direction to the Hadley cells: out towards the poles, and eastward, near the surface, giving us our prevailing westerlies; and towards the Equator and westward at higher levels of the troposphere.

E In the polar cells, at 50° to 60° south and north of the Equator, air sinks down towards Earth's surface at the poles and flows out towards the lower latitudes as it nears the surface.

EARTH'S SPHERES

It is only since the 1980s that modern science has begun to understand the inter-connectedness of Earth's systems and the complex network they form. Any event in one sphere can bring changes in other spheres. Events can be natural, like earthquakes or volcanoes, or result from human activity like oil spills. Small local changes can have unexpected worldwide consequences. Earth system science studies how all the spheres, systems, natural cycles and processes of Earth's environment work together.

Carbon and water move back and forth between different spheres, changing form, in their own constant cycles. Most of the carbon on Earth is locked in rocks, soil and sediments, some is stored in the ocean and the seabed, some in living organisms, especially plants, and the rest stays in the atmosphere in the form of carbon dioxide (CO_2), a greenhouse gas. For Earth's system to work smoothly, everything has to be in balance.

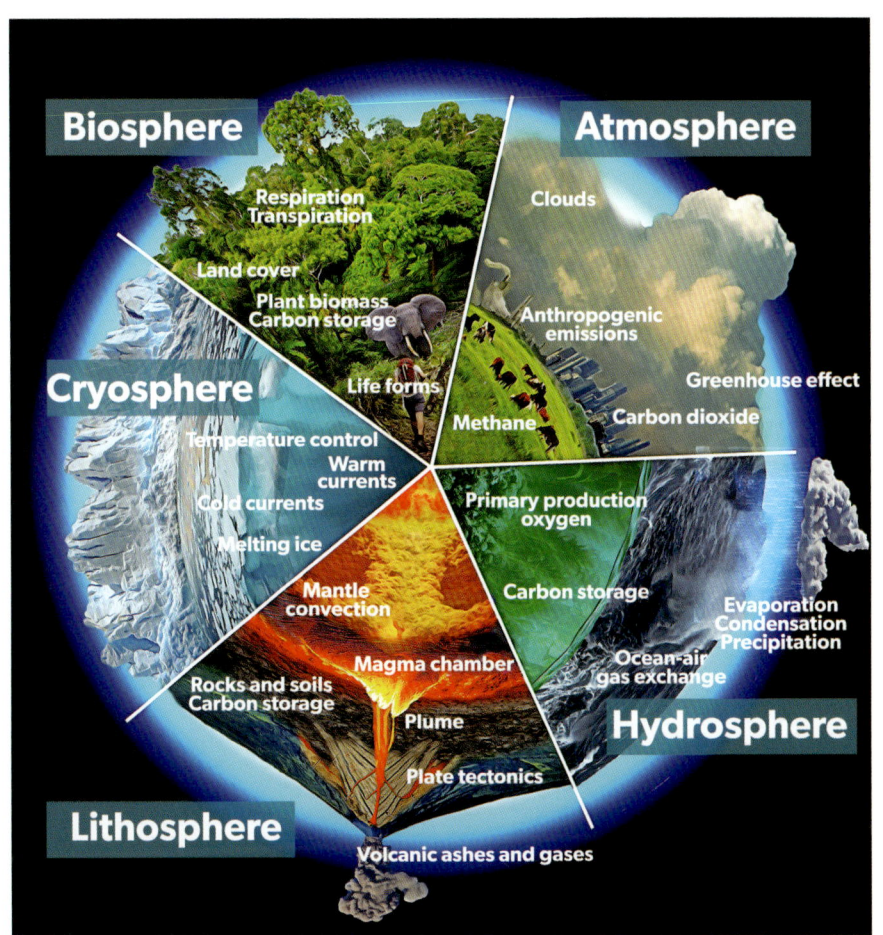

THE LITHOSPHERE The cold, hard rock of Earth's surface, and the hot, semi-solid and molten (liquid) rocks beneath Earth's crust.

THE BIOSPHERE All the living organisms on land, in air and sea, including humans, and the physical (non-living) components, such as temperature and light.

THE HYDROSPHERE All the water, fresh and salty, in all its forms — solid, liquid and gas — in its endless cycle of evaporation, condensation and precipitation: as water vapour in the clouds, on Earth's surface in lakes, rivers and seas, and beneath the surface as groundwater.

THE CRYOSPHERE All the frozen water, in ice sheets and glaciers. Over 70% of the world's surface is covered by water and about 97% of all water is in the oceans, which means only about 3% is fresh water and about 2% of that is found in ice caps and glaciers, which cover about 10% of our planet's land.

MATARIKI Traditional cultures have long recognised the interconnectedness of all things. In te ao Māori, (the Māori world) the key elements of the environment shine in the stars of Matariki. For some iwi the stars are a mother and her daughters. For others, the stars are paired, with male and female counterparts — their respective positions in the sky reflecting a world that is in balance.

Matariki, also known as the Pleiades or the Seven Sisters (though sometimes nine or more stars in this cluster are named), was one of the star clusters used by early Polynesian navigators to make their way across Te Moana-nui-a-Kiwa, the Pacific Ocean. There are about 500 stars in Matariki but only a handful of them are visible without a telescope. When the stars are clear and bright it means the coming growing season will be good. When they look blurry the growing season will be less fruitful. Matariki is not visible in the whole of New Zealand so for some iwi the rising of Puanga (Rigel in the Orion constellation) marks the beginning of the new year.

THE GOLDILOCKS PLANET

Without an atmosphere — the invisible, protective blanket of gases encircling Earth — our planet would be about 33°C colder than it is; a distinctly chilly -18°C. Our atmosphere makes Earth the 'Goldilocks planet', with temperature and conditions 'just right' for human habitation — not too hot and not too cold.

To keep our world's temperature and climate stable, the incoming energy from the Sun has to balance the energy Earth loses to space. Some gases in Earth's atmosphere, known as greenhouse gases, let the Sun's heat (solar radiation) pass through the atmosphere but prevent it from reflecting back into space. This is known as the natural greenhouse effect.

The strength of Earth's greenhouse effect depends on the concentration of the greenhouse gases in the atmosphere. The main ones are water vapour, carbon dioxide (CO_2), methane and nitrous oxide, which occur naturally, and man-made ones like the fluorinated gases used in fridges and air conditioners, which are far more potent than the natural gases. The greater the proportion of greenhouse gases in the atmosphere, the more of the Sun's heat is trapped.

EARTH'S HEAT BALANCE — THE NATURAL GREENHOUSE EFFECT

Only about half the Sun's rays reach Earth's surface, heating up the ground and the surface of the ocean. A smaller proportion of the Sun's heat is absorbed by clouds, water vapour, dust and gases in the atmosphere. The greenhouse gases help stop most of the heat that is emitted from the warmed ground and oceans from escaping into space. Some of the Sun's incoming energy is reflected back from the ground surface and some is scattered into space from clouds, dust and gases in the atmosphere, preventing our planet from heating up and keeping the temperature in perfect balance.

BALANCING ACT

Many factors can upset this delicate balance, from the amounts of greenhouse gases in the atmosphere, to how well different parts of Earth's surface reflect heat.

The concentration of CO_2 in the atmosphere is measured as the number of parts of CO_2 per million parts of a sample of air (parts per million/ppm). Records from Mauna Loa Volcanic Observatory in Hawai'i show that levels have risen steadily since the 1950s and the world's average surface temperature has risen in step over the same period. The graph below shows that the more extra greenhouse gases humans add to the atmosphere, the more of the Sun's heat is trapped and the more Earth warms.

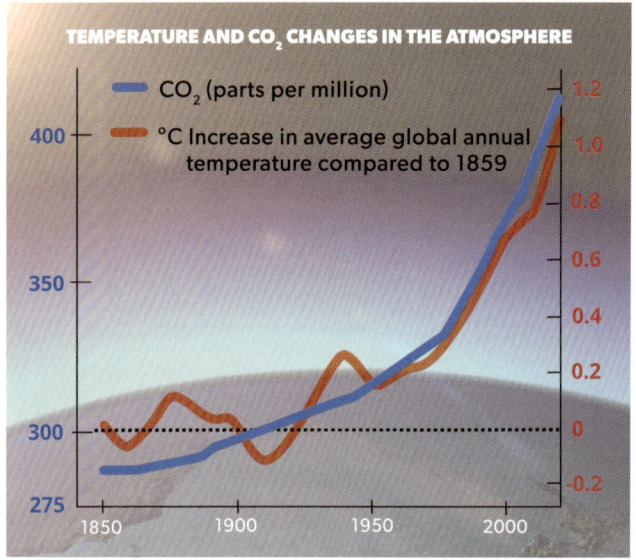

When the Industrial Revolution got started, around the mid-1700s, the concentration of CO_2 in the atmosphere was around 260–280 ppm. The safe limit, the limit to keep Earth's temperature balance at what we are used to, is thought to be 350 ppm.

ALBEDO

Different surfaces reflect sunlight to differing degrees. Ice and snow reflect most of the solar radiation that reaches them, which helps keep cooler regions cold. Darker surfaces like rocks, forests and oceans reflect very little of the Sun's energy and absorb it instead. The *reflectivity* of a surface (the amount of energy it reflects) is called *albedo*. Very dark colours have a low albedo and very light colours a high one. Changes in land and ocean cover, like the melting of ice sheets, affect the balance of light and dark surfaces and alter Earth's albedo and critical heat balance.

WEATHER WATCHERS

MetService's team of meteorologists, oceanographers, modellers and researchers analyse weather data from around the world and combine it with New Zealand measurements to produce forecasts. There are general predictions and specific predictions for towns and cities, national parks, rural and mountain areas as well as aviation and marine forecasts. Weather warnings are issued for heavy rain, snow and swell, thunderstorms and severe gales. MetService weather stations record many different aspects: barometric pressure, rainfall, soil moisture and temperature, outdoor temperature, visibility, wind direction and speed, outdoor humidity and dew point, sunshine hours, snow depths and even the minimum temperatures of grass and concrete. Weather enthusiasts can share their own readings on the Weather Observations Website (WOW) that MetService runs in conjunction with the UK Met Office.

The earliest New Zealand weather stations were in municipal gardens, harbours and lighthouses set up in the mid-1800s. The first weather forecasts were storm warnings for mariners, to help prevent shipwrecks. During World War 2 the Royal New Zealand Air Force took over weather forecasting and airfield control towers recorded weather data. In the 1980s Automatic Weather Stations (AWSs) began to replace staffed stations and the remote islands were no longer inhabited by meteorologists year round. Today, AWSs often include webcams and they are able to detect and distinguish hail and snow, rain and showers, as well as different cloud layers.

> Aotearoa's warmest year on record to date is 2021 and seven of the previous nine years were also record warmest years. The nationwide average temperature for the year was almost one degree (0.95°C) higher than the 1981–2010 annual average. (NIWA)

Meteorological station on Campbell Island, 700 km south of the South Island.

Since 1909 the National Institute of Water and Atmospheric Research (NIWA) has monitored changes in New Zealand's national average temperature at seven stations throughout Aotearoa: in Auckland, Wellington, Masterton, Nelson, Hokitika, Lincoln and Dunedin. The National Climate Database holds climate records from around 600 weather stations around New Zealand operated by NIWA, MetService and other providers, all contributing to the national database.

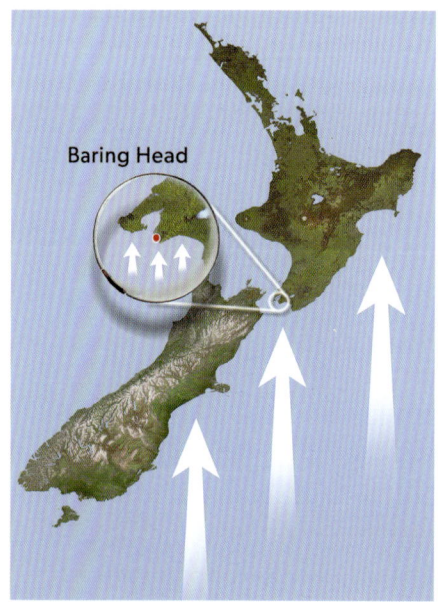

PURE AIR

Since 1972 atmospheric CO_2 levels have been measured and recorded at NIWA's clean air station at Baring Head, near Wellington. Baring Head is part of a global CO_2 monitoring network and the oldest station in the southern hemisphere. The location was chosen because, in a southerly wind, the air sampled there is as pure as it can be, having blown over nothing but ocean. Weather data is harder to come by in the southern hemisphere because there is more sea than land.

EARTH'S CLIMATE

Geologists divide Earth's history into eons, eras, periods, epochs and ages. They decide the start and end points of these spans of geological time from evidence of changes in layers of rocks (rock strata) and other fossilised materials. Volcanic eruptions, mass extinctions, asteroid strikes, 'explosions' of new life-forms and shifts in climate and vegetation all leave traces in the geological record.

Earth's climate has always changed. Over billions of years our planet has moved in and out of ice ages, and alternated between glacial phases (glaciations) when thick ice sheets cover the globe, and milder phases like the one we are in, called interglacials. We are currently in the Quaternary period, which began 2.6 million years ago, and in the Holocene epoch, an interglacial which started around 11,700 years ago.

Until recently (in the context of Earth's long history) the Holocene has been a relatively stable time with fairly steady temperatures, sea levels and concentrations of carbon dioxide in the atmosphere. But now that has changed and geologists believe we have entered a new epoch, the Anthropocene. The rate of change is much, much faster than usually seen in the climate record.

ERA	PERIOD	AGE (MYA)	EPOCH
CAINOZOIC (PHANEROZOIC)	Quaternary	01	Holocene
		2	Pleistocene
		5	Pliocene
	Tertiary	26	Miocene
		37	Oligocene
		53	Eocene
		65	Palaeocene
MESOZOIC (PHANEROZOIC)	Cretaceous	136	
	Jurassic	190	
	Triassic	225	
PALAEOZOIC (PHANEROZOIC)	Permian	280	
	Carboniferous	320	
	Devonian	345	
	Silurian	395	
	Ordovician	430	
	Cambrian	500	
		570	
PRECAMBRIAN	Proterozoic	2300	
		2800	
	Archean	4600	
		4700	

ANCIENT CLIMATE

Palaeoclimatologists study ancient climate — Earth's climate in the times before weather was measured using instruments and records were kept. They use proxy records of temperature, humidity and rainfall found in fossilised natural materials. They analyse physical, chemical and biological signatures left in layers of rock and sediment, and in seasonal and annual growth rings of corals, ice and trees, to study the conditions at the time these were formed and how Earth's climate changes. The dates of known events from more recent history, like volcanic eruptions and severe droughts or floods, are matched with the evidence in the proxies, to better understand the changing cycles of the world's climate.

The dark ring in this ice core is ash from a volcanic eruption. Annual growth rings like this tie a sample to a specific year.

NATURAL CLIMATE CHANGES

When Earth's climate shifts, changes in the lithosphere, hydrosphere, atmosphere, cryosphere and biosphere all play interacting roles, setting off complex webs of changes between and within the 'spheres'. Proxy records show Earth's movement in and out of ice ages accompanied by tectonic plate activity as continents broke apart and shifted position, and by earthquakes and volcanoes and changing levels of CO_2 in the air. These can magnify or reduce the different effects, causing both positive and negative feedbacks and speeding up or slowing down climate change.

Palaeoclimate records show that sea levels and CO_2 concentrations fall as Earth moves into an ice age and rise as the planet moves into warmer periods. In past climate shifts the movements of the plates in the Earth's crust brought earthquakes and volcanic eruptions, which altered the seabed and disrupted ocean circulation patterns. Changing ocean currents affected the exchange of heat between the atmosphere and the oceans and the workings of the water and carbon cycles too.

As glaciers melted, more solar radiation was absorbed by the newly exposed dark surface; temperatures rose, CO_2 levels slowly built up in the atmosphere, sea levels rose and the Earth gradually moved into an interglacial phase, like our current one.

Volcanoes hurl debris and gases into the atmosphere, which can block the Sun and cool the planet. The same additional greenhouse gases can have a warming effect. Volcanic eruptions disrupt weather systems and lead to unusual weather, both heating and cooling, sometimes as much as 4°C, in different parts of the world.

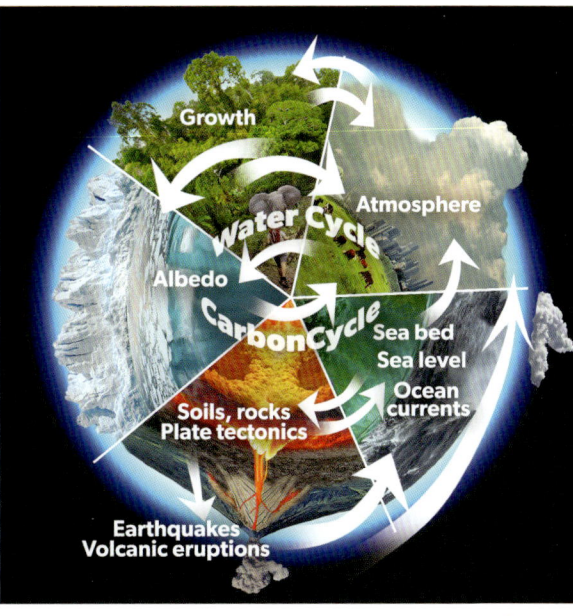

Increases in plant and forest cover and growth cool the planet by absorbing and storing CO_2. Wildfires and deforestation lead to increased heating.

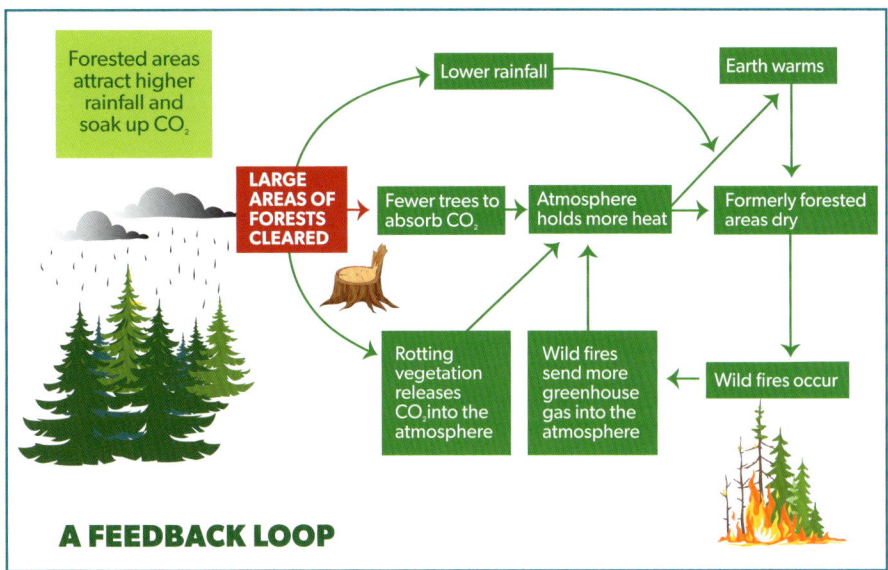

A FEEDBACK LOOP

TIPPING POINTS

When a feedback loop spirals out of control it can lead to a tipping point. Initial changes can seem minor but once a tipping point is reached things happen fast.

A tiny change could tip any one of these elements of the Earth system into a completely different state. They could tip one another over in a *tipping cascade* and push Earth beyond the point at which it could be stabilised, making it unliveable.

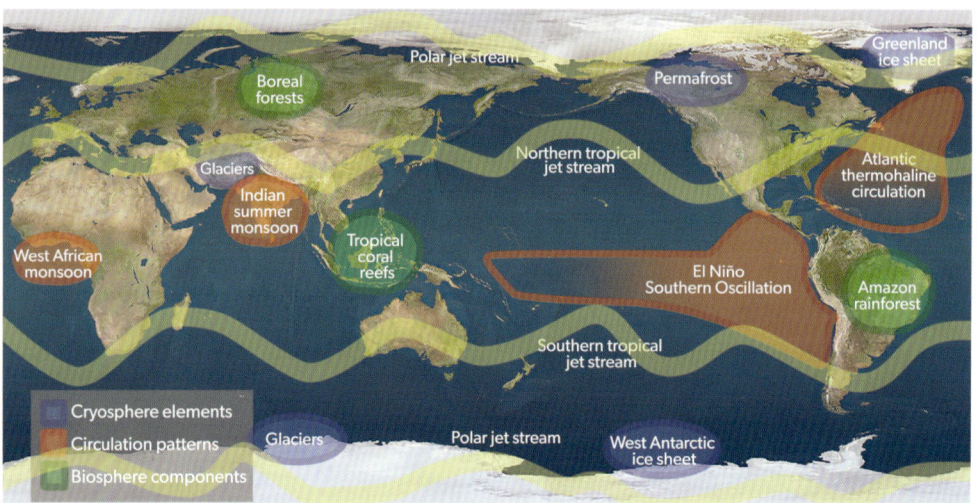

NATURAL DRIVERS OF CLIMATE CHANGE

Sunspots

Natural processes such as variations in the Sun's output and in our planet's orbit around the Sun play a role in changes in Earth's climate.

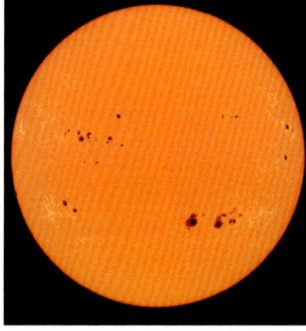

Sunspots are areas of concentrated magnetic fields on the Sun's surface, which may be associated with a slight lowering of energy output. They are cooler than the surrounding areas, so look darker.

These natural *forcings*, as they are called, appear to go in cycles — different cycles of very different lengths. Sunspot phases have a very small impact on Earth's temperatures. Even so, their contribution is included in climate models.

THE MILANKOVITCH CYCLES

The amount of sunlight that reaches Earth varies naturally, depending on how far our planet is from the Sun. A century ago Serbian scientist Milutin Milanković described three separate natural cycles in Earth's orbit of the Sun that can act as climate forcings. The three cycles involve:

- the tilt of Earth's axis (*axial tilt* or *obliquity*), relative to its plane of orbit round the Sun.
- how much Earth wobbles on its axis as it spins (*precession*).
- the shape of Earth's orbit round the Sun (*orbital eccentricity*). Earth's orbit is elliptical (much closer to an oval shape than circular) so, over the 100,000 year-cycle, sometimes the Sun is much closer to Earth than at other times.

The resulting changes in the solar energy reaching Earth do not affect the whole planet equally. Different latitudes receive varying amounts of sunlight over different months of the year. This is one of the reasons why heating and cooling occurs unevenly over the globe.

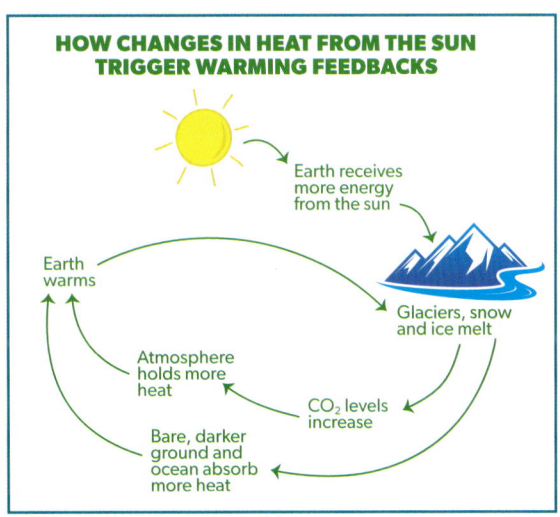

SNOWBALL EARTH

For much of the Cryogenian period (850–635 million years ago) Earth was mostly covered in ice. This phase is often called 'Snowball Earth'. Scientists think that when the supercontinent Rodinia broke up, a series of massive volcanic eruptions caused a build-up of gases and particles in the atmosphere that blocked sunlight. This, along with changes in ocean currents, contributed to a runaway climate effect and the sudden cooling of the planet. Geologic records show that Earth moved quickly out of the Snowball Earth phase and, in an abrupt climate reversal, into a very hot period, and then back into another snowball episode.

CLIMATE MODELLING

Supercomputers are used to run climate models using mathematical equations to represent the processes and interactions that drive Earth's climate. The models help scientists understand how climate works and how it may change in the future.

Today, integrated assesment models (IAMs) incorporate aspects of human societies like population figures, economic growth and energy use, since these all impact the Earth system. Today's models take into account forcings such as the amount of aerosols (minute airborne particles) from forest fires or volcanic eruptions, as well as climate feedbacks like changing albedo caused by melting ice.

Model predictions from 20 years ago have been matched with actual changes in temperature, and those early basic models have proved accurate. The climate models used in Intergovernmental Panel on Climate Change (IPCC) reports are run with different combinations of forcings to produce an array of possible futures. Nowadays they include 'attribution', assessing how much of climate change is due to natural influences and how much to human activities.

Combining global climate models with local data, NIWA's interactive website 'Our Future Climate New Zealand' shows projections for national and regional temperature and rainfall figures (among other aspects of climate) in the near future and over different timeframes to the end of the twenty-first century.

These scenarios range from continuing emitting the same or even higher amounts of greenhouse gases ('business as usual') to rapid and substantial emissions reduction, with two intermediary stages. If we do nothing to radically change how we live, parts of New Zealand could see temperatures 3–4 degrees higher before the end of this century. Such large increases will not only bring more frequent extreme weather events but will generally disrupt our way of life.

Rising temperatures are expected to bring more droughts and wildfires; especially to the north and east of New Zealand. On 4 October 2020, at Lake Ohau in the South Island, a fire burned 5000 hectares of land, with 48 homes destroyed. It took nine days for fire fighters to fully extinguish the fire.

The fire at Lake Ohau destroyed many buildings.

CARBON

What is carbon neutral?
Carbon neutral is when the amount of CO_2 released into the atmosphere through an activity balances the amount of CO_2 absorbed or removed from the atmosphere. This can also be called net zero carbon emissions or net zero. CO_2 makes up about 76% of global greenhouse gas emissions. Other gases like methane, nitrous oxide and CFCs make up around 24%. Carbon can be removed from the atmosphere by natural methods, such as increased planting of trees or by new technologies such as carbon capture and storage.

What is a carbon footprint?
Your carbon footprint is the total amount of greenhouse gases (including CO_2 and methane) that is added to the atmosphere by your actions.

The carbon budget
Between 1850 and 2019 human activities emitted 2400 billion tonnes of CO_2 into the atmosphere. In 2021 we were emitting about 40 billion tonnes per year. Every 100 billion tonnes is calculated as likely to cause half a degree of extra warming. At 2021 emission rates it was estimated that we could only emit 300 billion tonnes of CO_2 if we wanted to restrict warming to 1.5°C above 1850–1900 levels. So, in 2021, if we continued emissions at the same rates, we had less than eight years' worth of carbon to go.

> **YOUTH ACTIVISM AGAINST CLIMATE CHANGE** In 2018, at age 15, Swedish climate activist Greta Thunberg sat alone outside the Swedish parliament, calling for a radical reduction in greenhouse gas emissions. Since then she has rallied roughly 10 million people from 260 countries to join her School Strike 4 Climate campaign. In 2019 more than 20,000 students in New Zealand joined the protest and in 2021 thousands marched again.

METHANE
Methane is currently responsible for about half a degree of total warming. Over the course of a hundred years, methane is at least 25 times more potent a greenhouse gas than CO_2 but it lives in the atmosphere for about 12 years. Cutting down methane emissions is the fastest way to limit global heating. In the 2021 Glasgow Climate Pact, New Zealand agreed to cut methane emissions by 30% from 2020 levels by 2030. Most of New Zealand's methane emissions come from agriculture (see pages 21, 98).

'CODE RED FOR HUMANITY'
The 6th IPCC assessment report (AR6) was released in 2021, just as the effects of 1.1°C of additional heating were being felt. There

> **WHY 1.5 DEGREES?** It is now impossible for us to limit extra warming to 1.3 or 1.4 degrees. Greenhouse gases persist in the atmosphere for so long that continued warming is locked in to the climate system for another 20–30 years, even if we cease all emissions immediately. In 2015 the target was to limit global warming to less than 2 degrees above pre-industrial levels. By mid-2021 it was clear that every tenth of a degree of temperature rise makes a huge difference to the frequency and intensity of extreme weather events. At the beginning of 2022 the lowest achievable amount of heating was 1.5 degrees.

were deadly heatwaves and wildfires in Canada, North America and Siberia, and destructive flooding across Europe and Southeast Asia, with many hundreds of lives lost. New Zealand experienced the hottest June since records began, as well as intense rainfall and flooding in Canterbury and on the West Coast of the South Island, fortunately with no loss of life.

UN General Secretary Antonio Guterres announced that AR6 represented a 'Code Red for Humanity'. In summary, the report's findings were:

- We will exceed 1.5–2°C of warming of the average global surface temperature this century, unless we drastically cut our emissions of CO_2 and other greenhouse gases.
- Every additional half a degree increases the frequency and intensity of weather extremes.
- We need to reduce the use of fossil fuels and use more wind and solar and other renewables to reach net zero as soon as possible.
- Warming would have been half a degree greater if it were not for pollution from transport and industry reducing the amount of sunlight reaching Earth's surface. Ironically this means temperatures could spike even higher as we cut fossil fuel use.
- In 2022 Guterres reiterated his call for action. He said, 'We are on a fast track to climate disaster' and 'an unliveable world', stating that greenhouse gas emissions must start to fall by 2025.

MENDING THE PLANET

It is in our power to recognise environmental harm we have caused to the planet and take steps to mend the damage. We did this before, in the 1970s, when acid rain caused by pollution from burning coal destroyed swathes of northern forest. The story of the ozone hole shows how, by working together, we can restore balance to Earth's systems that have been upset by human actions.

OZONE

- Ozone absorbs most solar ultraviolet radiation and helps shield us from its harmful effects.
- The ozone layer is in the stratosphere, about 15–50 km above Earth's surface.
- Natural ozone forms when the Sun's ultraviolet radiation interacts with oxygen in the atmosphere.
- Without sunlight, ozone levels in the upper atmosphere above the poles fall in wintertime.

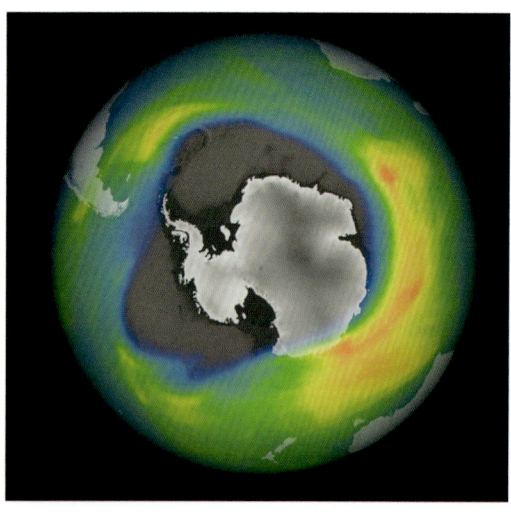

When the ozone layer is depleted (thinned) more of the Sun's harmful ultraviolet rays reach Earth's surface. Harmful UV rays can damage DNA in cells, damage plants and marine ecosystems and cause eye problems, and sunburn that potentially leads to skin cancer. New Zealand has one of the highest rates of skin cancer and melanoma in the world. This is in part due to the clarity of the air in the skies above Aotearoa and partly due to the hole in the ozone layer over Antarctica.

THE OZONE HOLE

Ozone is created mainly above the sunny tropics and the ozone-rich air flows out towards the poles. It has a yearly cycle and varies with the seasons and changing air pressure. In the dark winter the polar vortex, a swirling band of low pressure and strong winds that circulates high above the pole, keeps the cold, ozone-poor air above Antarctica. When sunlight returns to polar regions in the spring, chlorine and bromine molecules from CFCs and halons break down the ozone and the 'hole' in the ozone layer grows.

The size of the ozone hole varies from year to year but it is at its largest over Antarctica in early spring. As the polar vortex weakens with the lighter days and warmer weather, ozone begins to rebuild, but chlorine depletes ozone faster than it can be replaced.

> **CFCS** CFCs (*chlorofluorocarbons*) and *halons* are artificial chemicals used in refrigerants, aerosol sprays, foam insulation and fire extinguishers. They hold tens of thousands of times more heat than molecules of carbon dioxide and can persist in the atmosphere for more than a century.

When the ozone hole grows the winds that circulate around the South Pole strengthen and air pressure over New Zealand increases. This can trigger weaker westerly winds in the spring, with less rainfall on the west coast of the South Island. Sometimes the upper atmosphere circulation systems shift, exposing New Zealand to ozone-poor air and greater levels of damaging ultraviolet light.

FIXING THE HOLE

In 1987, many years after the negative effects of CFCs on the ozone layer were first reported, nearly 200 countries signed the Montreal Protocol, pledging to phase out these harmful gases. As a result, the ozone layer is now gradually recovering and is expected to be back to 1980s levels around the middle of this century.

Ozone levels are monitored by satellite instruments, weather balloons and LIDAR (light detection and ranging). In Aotearoa, CFC and ozone levels are tracked and measured at the NIWA Atmospheric Research Station at Lauder in Central Otago. The research station is part of the international Network for the Detection of Atmospheric Composition Change (NDACC), which monitors trace gases linked to both ozone depletion and global warming. An ozonesonde, fixed to a weather balloon, samples air at regular intervals as it rises through the atmosphere and radios measurements back to base at Lauder and a Dobson spectrophotometer measures ozone concentrations from the ground. NIWA produces ozone maps from the readings.

> **THE SUN** The corona is the brightly shining area you can see around the Sun in a solar eclipse. It constantly changes size and shape, affected by the Sun's magnetic field. The energy the Sun emits is also constantly changing. It is made of plasma, a kind of superheated, electrically charged gas, with a denser core that generates nuclear reactions, producing the heat and light we receive on Earth. The churning movements of the plasma produce powerful magnetic fields, twice the strength of Earth's fields.
> Coronal mass ejections (CMEs) are explosive bursts of plasma from the corona that take anything from 15 hours to several days to reach Earth.

OZONE AND ULTRAVIOLET (UV)

NIWA measures the amount of UV (ultraviolet) radiation from the Sun and combines this with the ozone data to give daily forecasts of the UV index. This lets people know how strong the Sun is and how high the risk of skin damage. The scale ranges from 1, indicating low risk, up to 11+. The values for the South Island are generally lower than for the North Island, as UV is higher in the skies above the tropics. The higher the Sun and the clearer the sky (of both cloud and pollution), the higher the UV index is likely to be. There are monitoring stations at dozens of locations around New Zealand, including at ski fields.

Of all the radiation that streams from the Sun, we are only able to see visible light — a very narrow range in the middle of the electromagnetic spectrum. Ultraviolet light (*ultra* is Latin for beyond) is just outside our range of vision. Most of the Sun's energy output is in the infrared, visible and ultraviolet parts of the spectrum. X-rays come only from the corona, the outer layer of the Sun's atmosphere. Gamma rays are generated in the Sun's core and only reach Earth in the event of the most powerful solar flares. Gamma rays, X-rays and some ultraviolet rays have such high energy that they are ionising; in other words, they create ions (electrically charged particles) by jolting electrons out of atoms.

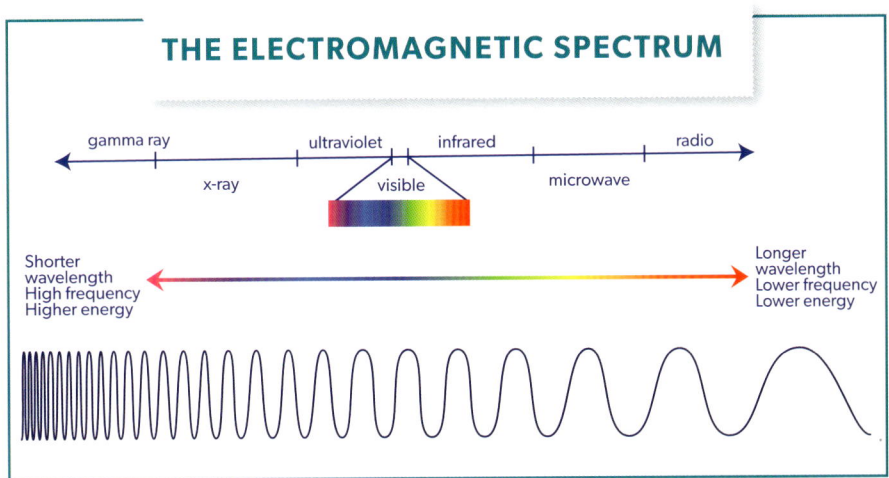

THE SUN AND SPACE WEATHER

The solar wind is the flow of electrically charged particles and magnetic clouds that streams out non-stop from the Sun's churning corona into space (see p. 36). When the particle radiation, magnetic fields and clouds of subatomic particles and other matter that explode from the Sun's surface interact with Earth's magnetic field and upper atmosphere they can cause geomagnetic storms, also known as space weather.

> **SOLAR FLARES** are sudden intense bursts of magnetic energy that flare up brightly from the Sun's surface, heating to millions of degrees. They release electromagnetic radiation, including X-rays, ultraviolet radiation, visible light and radio waves, travelling at the speed of light, reaching Earth's atmosphere in 8½ minutes.

EARTH'S MAGNETOSPHERE

Earth is like a giant magnet. It has magnetic poles and a magnetic field generated by the motion of molten iron in its core. The magnetosphere is the ever-shifting zone of magnetic fields and trapped charged particles that interacts with the solar wind.

Earth's magnetic field stretches tens of thousands of kilometres into space. It protects us from most solar and cosmic particle radiation but, in a geomagnetic storm, solar wind particles travel through magnetic field lines to Earth's atmosphere above the poles.

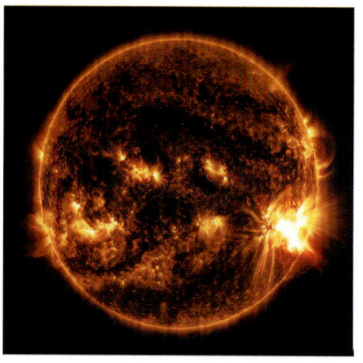

A magnetic storm brings a sudden drop in the strength of our planet's magnetic field that can last for many hours, disrupt high frequency radio communication and satellite navigation systems, disable satellites and knock out electrical power on Earth.

THE AURORA

Auroras are caused by the bursts of energy released by the collision of particles in Earth's atmosphere above the poles during geomagnetic storms. In severe space storms they can even be visible in the tropics. Auroras occur about 100–400 km up and can extend thousands of kilometres. Different molecules glow different colours at different heights in the sky. Occasionally they have been reported as far north as Auckland. The NIWA research station at Lauder was set up in the 1960s to study the aurora.

Auroras can be an indicator of geomagnetic storm conditions. In times of high activity jet aircraft can be rerouted away from the poles and spacecraft instruments shut down until storm conditions pass and Earth's magnetic field recovers. In February 2022, around 40 newly launched satellites orbiting 210 km above Earth were knocked out by a geomagnetic storm.

THE IONOSPHERE

The ionosphere is part of the magnetosphere and is created by the Sun's ionising radiation. It forms in different regions of the mesosphere and thermosphere, between 60 and 1000 km above Earth's surface. Ions are electrically charged atoms or molecules that have been stripped of one or more of their outer electrons when they collide with the Sun's high-energy radiation. The ionosphere is used for shortwave radio communication which can be disrupted by space weather events.

The Aurora Australis, or Southern Lights, is readily seen in the South Island because they are closer to the South Pole. This stunning image was taken from a vantage point above Dunedin.

OCEANS AND CLIMATE

Aotearoa has more than 15,000 km of coastline and no place in the country is further than 130 km from the ocean. New Zealand's Economic Exclusion Zone (EEZ) — the area of sea and seabed around us over which we have rights of use — is by international standards vast for the size of the country. Our marine environment and resources are extremely important to us.

THE OCEAN CURRENTS: THE THERMOHALINE CIRCULATION

The oceans of the world interact with the atmosphere, absorbing solar energy and releasing heat. Ocean currents distribute heat and energy throughout the planet's oceans in a global conveyor-belt-like system called the Thermohaline Circulation or Meridional Overturning Circulation.

Oceans play a vital part in regulating the climate as they store so much more heat than the atmosphere. The ocean has absorbed

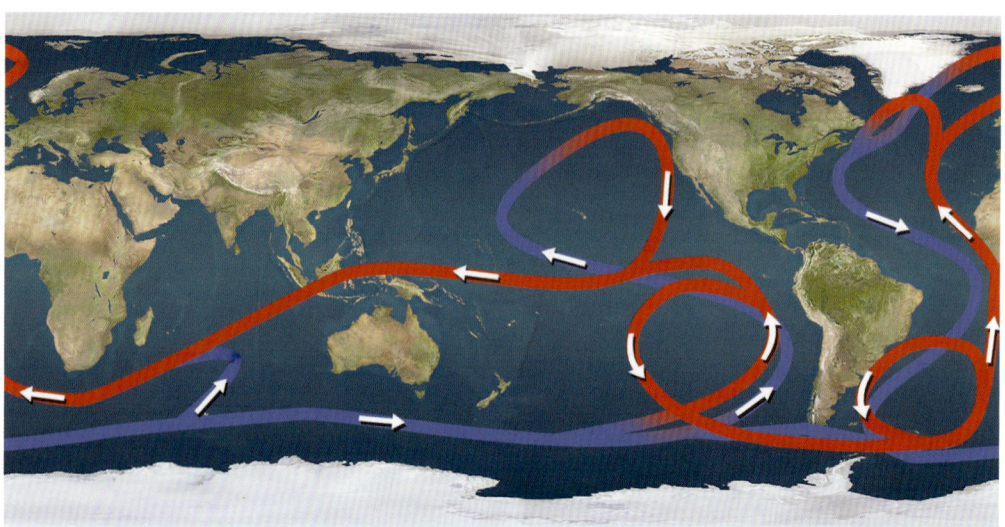

THE OCEAN CURRENTS: THE THERMOHALINE CIRCULATION

over 90% of the extra heat that human activities have put into the climate system, reducing the amount of extra warming in the atmosphere and softening the effects of climate change. However, this has resulted in an increase in the ocean's temperature. It was once thought that it was only the surface waters that were heating up but it is now clear that warming is also occurring in the ocean depths.

ARGO FLOATS The temperature of the oceans is monitored using a global network of Argo floats. Thousands of these floats drift freely on ocean currents. They rise to the surface at programmed intervals, measuring the temperature and salinity (saltiness) of the upper 2000 metres of the world's oceans as they go, and transmitting this data, along with the float's position, to satellites.

OCEAN CIRCULATION

Ocean circulation helps regulate climate by transporting and storing heat, carbon, nutrients and water around the globe. The movement of surface waters is driven by the wind and the much slower deep-sea circulation is powered by convection currents, driven in part by changes in the density of the seawater.

As the oceans warm, evaporation from the surface waters speeds up, increasing the amount of water vapour in the atmosphere. This supercharges the water cycle, adding more cloud to the skies, more energy for storms and cyclones and bringing torrential rainfall to some parts of the globe and drought to others.

Rainwater and melting ice make seawater less salty. This makes ocean water lighter, slowing the sinking process, which slows down the whole circulation system. The impacts of this are felt in different ways the whole world over; oceans are less able to absorb heat and carbon, leading to higher levels of both in the atmosphere. This can speed up global warming.

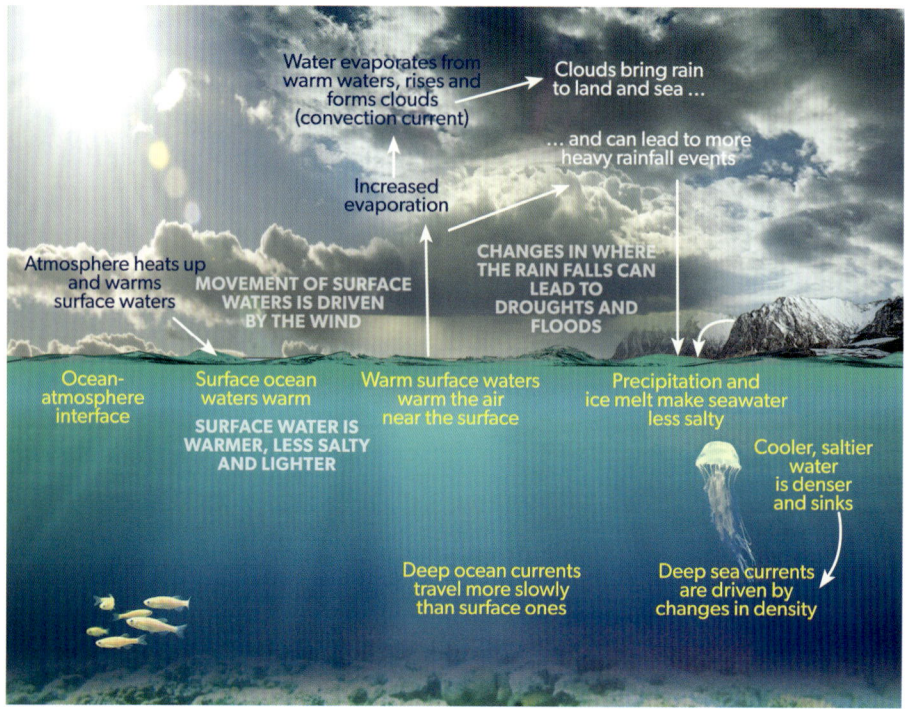

OCEAN-ATMOSPHERE INTERACTION

The oceans play a key role in the climate system by absorbing carbon dioxide from the atmosphere. Carbon dioxide dissolves in surface ocean waters and is then carried by ocean currents to the depths. CO_2 dissolves more easily in colder waters so the deep waters of the polar regions are important carbon sinks. As the seas warm, the amount of CO_2 that oceans can absorb is reduced. This is likely to further increase atmospheric warming.

As the oceans heat up, some tropical marine species are migrating to higher latitudes and fish not previously seen in New Zealand waters are appearing in our seas. Meanwhile, species that prefer cooler waters are moving towards polar regions.

> **WHAT IS A CARBON SINK?**
> A carbon sink absorbs and stores carbon from the atmosphere by physical or biological means. It can be natural or artificial; the key is it absorbs more atmospheric carbon than it releases. Forests, soils, limestone and oceans are natural carbon sinks. So are coal, oil and natural gas but when these and wood are used as fuels, the carbon they once stored is reinjected into the atmosphere and they become a carbon source.

SEA TREES 'Manawa' is Māori for mangroves, and means *heart* or *the breath*. Mangroves are the heart and lungs of the coastal environment. With their snorkel-like breathing roots, they grow inside harbours and river estuaries around Northland and Auckland and down to Raglan and the Bay of Plenty. They are nurseries for young fish and other kaimoana (seafood). They protect the shoreline from storm surges and prevent erosion. In times of heavy rainfall and flooding they absorb storm water and protect our waterways by trapping sediment and pollutants. Most of all, mangroves and seagrass meadows are a valuable carbon sink, even more efficient, hectare for hectare, at storing carbon than mature forest on land.

OCEAN ACIDIFICATION

Ocean acidification is the increase in the acidity of seawater. In the last 200 years it has increased by more than a quarter. The present rate of acidification is ten times faster than at any other time in the past 55 million years; this change in ocean chemistry is too fast for many marine organisms to adapt to.

New Zealand species such as mussels, oysters, paua and kina are all sensitive to the effects of rising levels of ocean acidity.

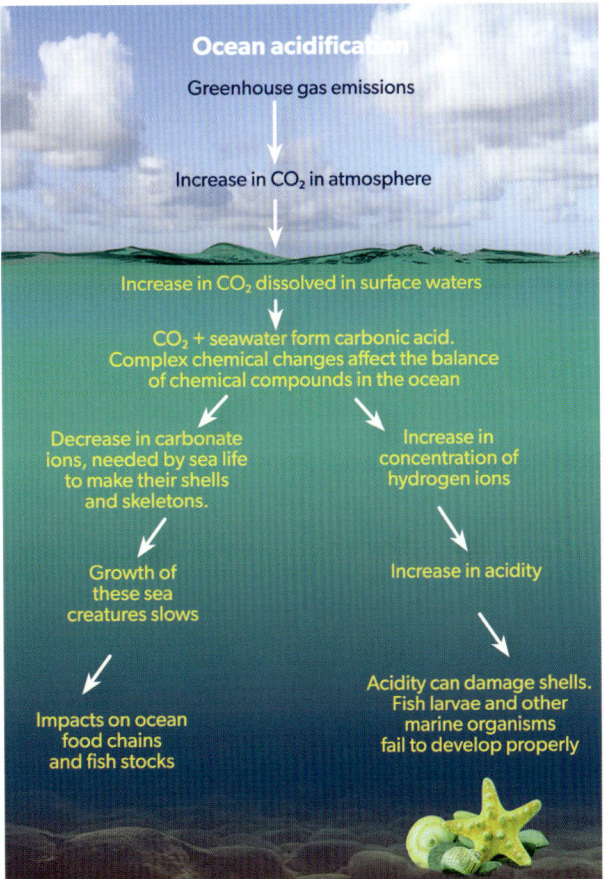

THE OCEAN AND CARBON CYCLE

All life on Earth is made of carbon, including humans. The total amount of carbon in the Earth system does not change but it changes form and continually moves between Earth's different spheres in what is called the carbon cycle. Most of the planet's carbon is stored in the lithosphere and deep beneath Earth's crust. As CO_2, it moves from the atmosphere into living organisms in the sea and on land and back again in an endless cycle.

The ocean is an especially valuable carbon sink. In 2021 more than 100 countries around the world agreed to protect at least 30% of their ocean territory by 2030 in the Glasgow Climate Pact.

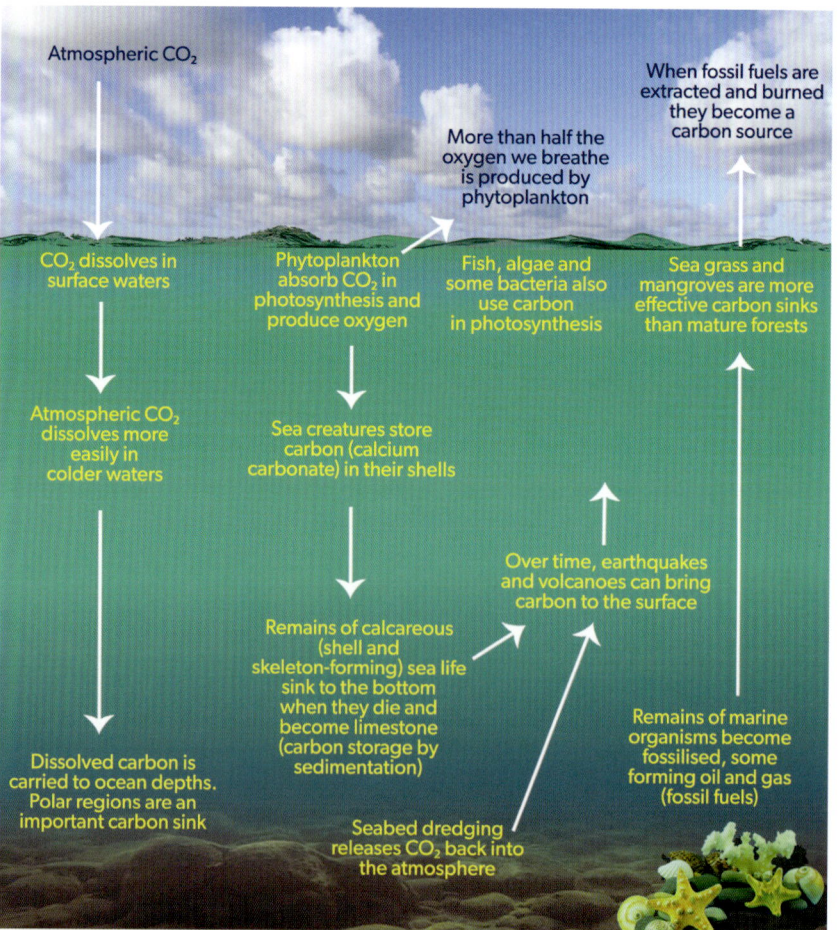

CHANGING SEA LEVELS

Sea levels are rising in two ways. Water expands as it heats so the volume of water in the world's oceans is expanding as the oceans warm. Sea levels are also rising because of the growing volume of water from melting glaciers. The average global sea level is projected to rise by up to one, possibly two metres by the end of this century and the 2021 IPCC report warned that a rise of five metres by the year 2150 cannot be ruled out.

Sea levels rise unevenly over the planet because of the varying effects of local winds and ocean currents. Some of Aotearoa's Pacific island neighbours are among the people at most immediate risk of losing their homes. Low-lying atolls are just a few metres above sea level and they are already suffering from storm surges and high tides, which destroy their crops, taint their drinking water, damage infrastructure and eat away their limited land.

New Zealand's coastline will be subject to more frequent and more extreme flooding and erosion as temperatures rise. Every centimetre of sea level rise means that storm surges start from a higher base and waves push further inland. If a storm makes landfall during a high tide,

Visit www.searise.nz to see how sea levels are expected to rise around Aotearoa's coastline.

damage to infrastructure such as buildings, roads and bridges can be all the worse. In many parts of the country beaches are eroding, cliffs are crumbling into the sea and people driven to leave their homes and move away from areas at risk of slips and flooding.

> **WHAT IS A STORM SURGE?** A storm surge is a double whammy; a combination of wind setup and the inverted (or inverse) barometer effect. Wind setup is when strong onshore winds push surface ocean water towards the land and cause coastal seas to build up higher than usual. The inverted barometer effect is when the sea surface responds to changes in air pressure. The sea level beneath a *low* pressure system like a storm or a cyclone actually *rises*. Similarly, in step with the amount of pressure change, it *falls* beneath a high-pressure system. One hectopascal (hPa) is 1/1013 of one atmosphere of pressure, so the sea level rises or falls about 1 cm for a pressure change of one thousandth of one atmosphere. This rise or fall has a negligible effect in the deep ocean, but a rise of just a few centimetres can make a significant difference where the sea meets the land.

> **ATMOSPHERIC PRESSURE**, also known as barometric pressure because it is measured using a barometer, is an indicator of weather. At sea level, air pressure (the collective weight of air molecules pressing down on Earth's surface), normally ranges between 970 and 1050 hectopascals (hPa). The average air pressure is 1013 hPa. Differences in air pressure are a result of unequal heating and unequal pressures caused by air flow over the Earth's surface.

OCEAN DEOXYGENATION

The volume of the world's ocean water that has become anoxic (completely depleted of oxygen) has increased fourfold since the 1960s. Oxygen has been lost mainly from the upper 1000 metres of ocean where the greatest diversity and numbers of sea creatures live.

Even a slight reduction of oxygen levels can stress marine organisms and affect the balance of marine life. Warming seas, environmental pollution, changes in rainfall, ocean currents and in the mixing of ocean water can all lead to ocean deoxygenation. In 2018 more than 500 coastal sites around the world were identified as having hypoxic (low oxygen) waters, including part of the south coast of the South Island.

MARINE HEATWAVE WARNINGS

A marine heatwave is declared when the ocean temperature at a given location is in the top ten percent of temperatures recorded for that time of year for at least five days. Sea surface temperatures are monitored at ten sites around Aotearoa, ranging from Cape Reinga to Stewart Island and from Fiordland to the Chatham Islands.

In 2018 the highest level of heating was off the South Island; in the summer of 2019–20 a huge area of ocean east of the Chatham Islands heated to 4° above normal and in the summer of 2021–22 some parts of the sea off the west coast of the North Island were 4° warmer than usual. In conditions like these fish can spawn earlier than usual or move south in search of the temperature range they are used to. Marine heatwave warnings are useful to people such as aquaculture operators, to help them plan for warmer-than-usual waters, either by moving their stock or harvesting early.

2021 IPCC AR6 REPORT: OUR OCEANS

- Carbon dioxide emissions are the main driver of ocean acidification.
- The ocean surface has warmed less than the land; two-thirds of that warming has occurred in the last 50 years.
- The oceans are expected to continue to heat up until at least 2300 because the ocean circulation system moves more slowly than the atmospheric one.
- In the northern hemisphere the Greenland ice sheet is melting, largely because of warmer air temperatures. In Antarctica, the West Antarctic ice sheet is melting faster because of warming ocean waters.
- The average global sea level is projected to rise between 0.4 metres (if we achieve net zero by 2050) and 0.8 metres (at the highest emissions scenario) above the 1995–2014 average, but if Antarctica melts faster than expected we could see an additional one metre of sea level rise by 2100.

TROPICAL CYCLONES

Our varied and often fast-changing weather comes to us from all directions across the surrounding oceans. Westerly winds from over the Tasman Sea drop heavy rain in the west; biting southerlies bring snow and cold snaps from the Southern Ocean, while every now and then some of the wildest weather swoops in from the vast Pacific Ocean.

Tropical cyclones start out as depressions or Lows in the warm tropics to the north of New Zealand. They are fuelled by the heat of the ocean in the hotter months, from November to April, the official cyclone season, but they can occur outside this time.

HOW TROPICAL CYCLONES FORM

Cyclones are created when warmed air rises above the ocean surface, drawing in cooler air to replace it and forming a low-pressure area. The Earth's spin makes the air rotate and as the updraughts of warm air strengthen, a cyclone develops. The sea surface temperature needs to be at least 26°C for this to happen.

By the time a cyclone reaches the cooler seas around New Zealand it has lost strength and is downgraded to an ex-tropical cyclone. However, it can still bring widespread storm-force winds, torrential rain and high seas and flooding to both sides of both main islands.

Tropical cyclones generally travel westwards at first, with the trade winds, then towards the pole and eastwards. Their paths are influenced by surrounding weather systems and meteorologists chart their tracks to forecast where and when they are likely to make landfall. Sometimes, when tropical cyclones pass over areas of extra warm water as they cross the ocean, they can strengthen before they reach land. When a cyclone weakens, its track slows

Barometric pressure is lowest in the 'eye' or centre of a storm.

and the storm can stall over land, bringing even longer-lasting and intense rainfall and flooding.

Barometric pressure is lowest in the 'eye' or centre of a storm. The lower the air pressure, the more powerful the cyclone. The strongest wind and heaviest rainfall come from the area surrounding the eye, called the eye wall.

When a tropical cyclone weakens it loses its shape. The extreme winds and rain spread over a much wider area and can be just as destructive as when a storm passes quickly and powerfully. When ex-cyclones reach New Zealand the most torrential rainfall tends to come from the southern portion of the low, rather than the storm centre, and stretch for hundreds of kilometres.

TRACKING CYCLONES
Cyclone Gita

In February 2018, Cyclone Gita started out as a tropical disturbance near the Solomon Islands and developed into a tropical cyclone near Niue a week later. As it tracked across the South Pacific, Tonga and Sāmoa were especially badly hit. By the time it hit New Zealand it had been downgraded to an ex-tropical cyclone but still brought extremely heavy rain and damaging winds to many areas, from Taranaki in the North Island down to Canterbury in the south.

As it passed over the lower North Island and upper South Island it split into two systems. Many districts, to the east and west, declared a state of emergency and almost 200 schools and preschools were closed. Port Taranaki recorded 15-metre-high waves, sea walls in Wellington were damaged, and Kaikoura received over 200 millimetres of rain — over a quarter of its annual rainfall. Gita caused so much damage in so many parts of the South Pacific that its name was removed from the official list of names for cyclones and will never be used again in this part of the world.

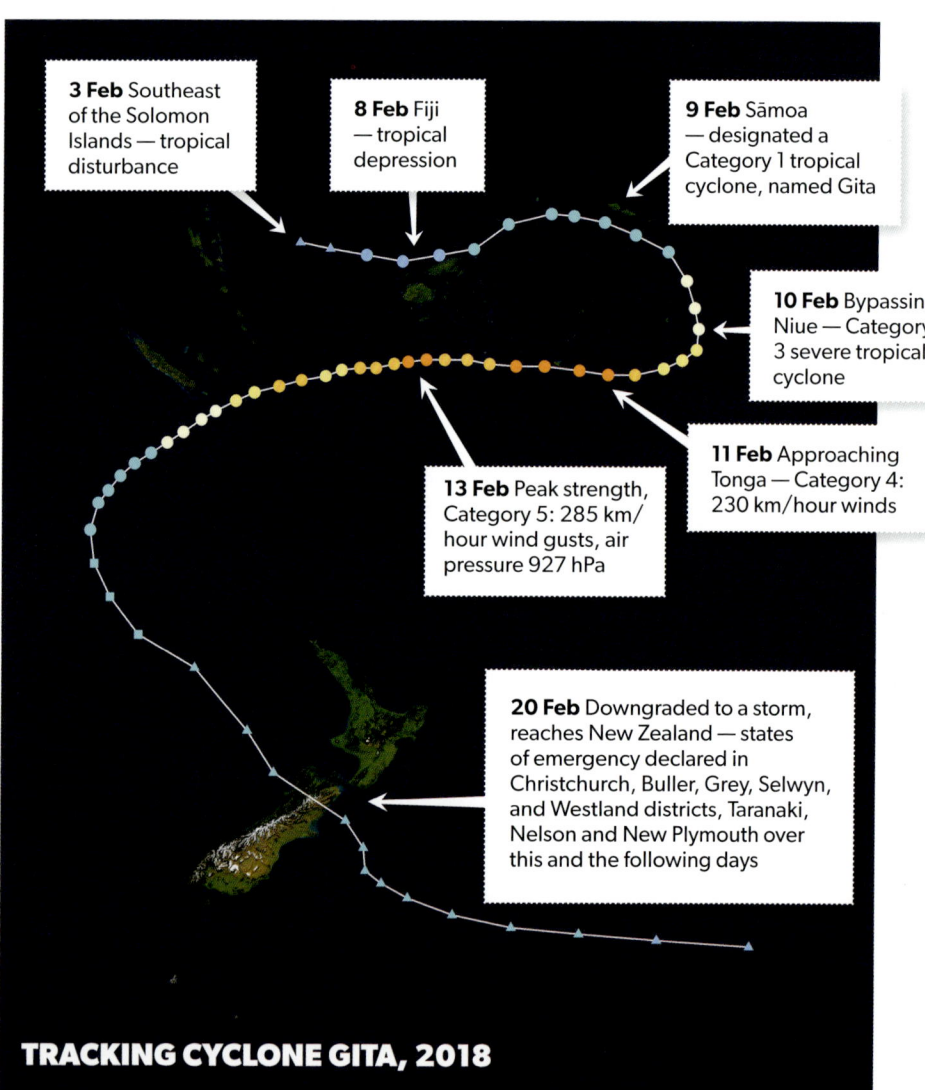

TRACKING CYCLONE GITA, 2018

- **3 Feb** Southeast of the Solomon Islands — tropical disturbance
- **8 Feb** Fiji — tropical depression
- **9 Feb** Sāmoa — designated a Category 1 tropical cyclone, named Gita
- **10 Feb** Bypassing Niue — Category 3 severe tropical cyclone
- **11 Feb** Approaching Tonga — Category 4: 230 km/hour winds
- **13 Feb** Peak strength, Category 5: 285 km/hour wind gusts, air pressure 927 hPa
- **20 Feb** Downgraded to a storm, reaches New Zealand — states of emergency declared in Christchurch, Buller, Grey, Selwyn, and Westland districts, Taranaki, Nelson and New Plymouth over this and the following days

CYCLONE FACTS

- Tropical cyclones can spread over many hundreds of kilometres. They are rated in intensity from 1 to 5. Category 5 is the most dangerous.
- Once the mean surface wind speed reaches 63 km/h for at least ten minutes, a tropical cyclone becomes a category 1. Category 3 (119 km/h winds) is a severe tropical cyclone and a category 5 is winds over 200 km/h, with gusts over 280 km/h.
- Each new season, tropical cyclones are named in alphabetical order from set lists used by each met service and starting with the letter A. Each cyclone is named by the met service of the region where it develops. From the New Zealand point of view, South Pacific tropical cyclones do not always seem to come in alphabetical order.
- Originally cyclones were always given women's names. Since 1979 names have alternated between male and female.
- Scientists study the tracks of past cyclones to help them predict the movements of active cyclones and where they are likely to make landfall.
- In El Niño seasons the tracks of tropical cyclones tend to loop in more irregular patterns, rather than follow a simple curving path. This makes it tricky to forecast their movements. It can also mean they impact a greater area.
- Tropical cyclones tend to move more slowly in La Niña conditions and faster in El Niño ones.
- Tropical cyclones do not usually develop in the region between 5° north and 5° south of the Equator where there is no Coriolis effect to generate a revolving storm.
- As temperatures rise, both in the seas and all over the Earth, tropical cyclones are developing more rapidly and becoming more powerful.

DOUBLE TROUBLE When an ex-tropical cyclone meets another powerful weather system the results can be all the more destructive. In April 1968 ex-Cyclone Giselle collided with a storm that came up the west coast of the South Island from Antarctica, bringing gusts of 270 km/h to Cook Strait. The interisland ferry *Wahine* sank in Wellington Harbour and 52 people lost their lives. The whole country was affected, from the Far North to the Deep South, with flooding, roofs ripped off houses, rescue vehicles blown over and roads blocked by landslips.

EL NIÑO SOUTHERN OSCILLATION

The El Niño Southern Oscillation (ENSO) climate cycle can affect the likelihood of ex-tropical cyclones reaching New Zealand shores. ENSO is an irregular pattern of weather phenomena that stems from complex ocean-atmosphere interactions in the South Pacific. At its heart is the Walker Circulation, the series of linked east-west cells or loops of circulating air that operate in a belt around the Equator. Of these, the Pacific Walker Cell, which transports air near surface level westwards across the South Pacific, has the greatest impact on world weather. Changes in the Walker Circulation and ENSO bring major shifts in wind, temperature and rainfall patterns to places as far apart as Africa, India, Southeast Asia and South America.

The Walker Circulation pattern can vary greatly in intensity from one season or year to another. This variation, also known as the Southern Oscillation, is fuelled by ocean surface temperature and air pressure differences between the eastern and western tropical South Pacific. Scientists refer to three phases of the Southern Oscillation: the neutral phase, El Niño and La Niña, all of which affect the weather of Aotearoa. In reality these are general categories and the changes in weather patterns that each brings are complex and vary because each El Niño and La Niña is different.

NEUTRAL CONDITIONS – THE WALKER CIRCULATION

A cold ocean current usually flows northwards from Antarctica up the west coast of South America, so the eastern side of the South Pacific is generally cooler than the west (**1**). The southeast trade winds push surface waters across the ocean (**2**) and warm water builds up on the western side, creating the Pacific Warm Pool (**3**). The warm, moist air above this warm water (**4**) rises high into the atmosphere, bringing unstable, stormy weather (**5**) and lower air pressure at the

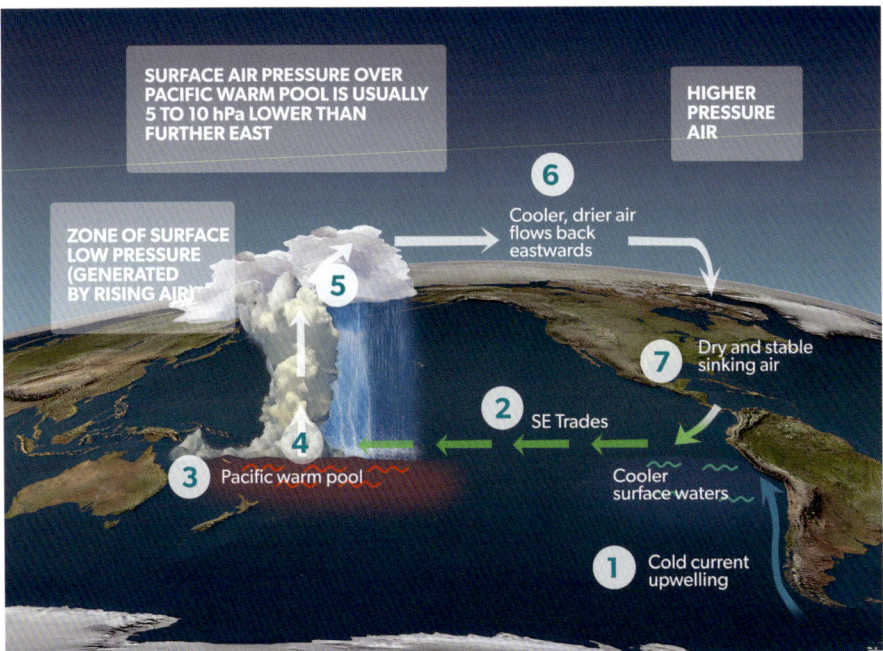

surface. High in the troposphere, the now cooler, drier air flows back eastwards (**6**) and then sinks back to the surface off the west coast of South America (**7**), completing the loop. This dry, stable, sinking air builds high surface pressure; the air pressure and temperature differences between east and west power the circulation cell.

THE SOUTHERN OSCILLATION INDEX (SOI)

In the western South Pacific, the surface air pressure in Darwin, Northern Territory, is usually 5–10 hPa lower than in Tahiti, French Polynesia, to the east. The Southern Oscillation Index (SOI) is a measure of the pressure difference between Darwin and Tahiti, and it indicates the strength of the Walker Circulation.

The index is high when the pressure difference between the two regions is at its greatest — when the Walker Circulation Cell is at its strongest. Every few years the pressure difference oscillates (swings) beyond the normal (neutral) range and we get either an El Niño, or its opposite, a La Niña phase. El Niño is declared when temperature rise is more than half a degree Celsius above average in the Eastern Pacific.

EL NIÑO

In El Niño conditions the southeast trade winds weaken or even reverse. The Humboldt Current, the upwelling of cold water along the South American Pacific coast, slows, (**1**) bringing an increase in sea surface temperature (**2**). Because it occurs around Christmas time, this unusual warming off the coast of Ecuador and Peru was called 'el Niño' ('the boy child' in Spanish) by local people, referring to the baby Jesus. Now the name has come to mean the large-scale weather conditions associated with the 'weak' or 'warm' phase of the Southern Oscillation, when the trade winds weaken. La Niña, which has opposite weather effects, means 'the girl child'. In an El Niño phase of ENSO the Western Pacific cools (**3**), the Pacific Warm Pool moves towards the east (**4**) and the sea level in the Eastern Pacific can rise by as much as 20 centimetres.

> **HOW EL NIÑO AFFECTS AOTEAROA WEATHER**
> Eastern areas are drier than usual • In summer the westerlies are stronger and more frequent, bringing more rain to the west • In spring and autumn southwesterly winds are stronger and more frequent • In winter there are more cold southerlies.

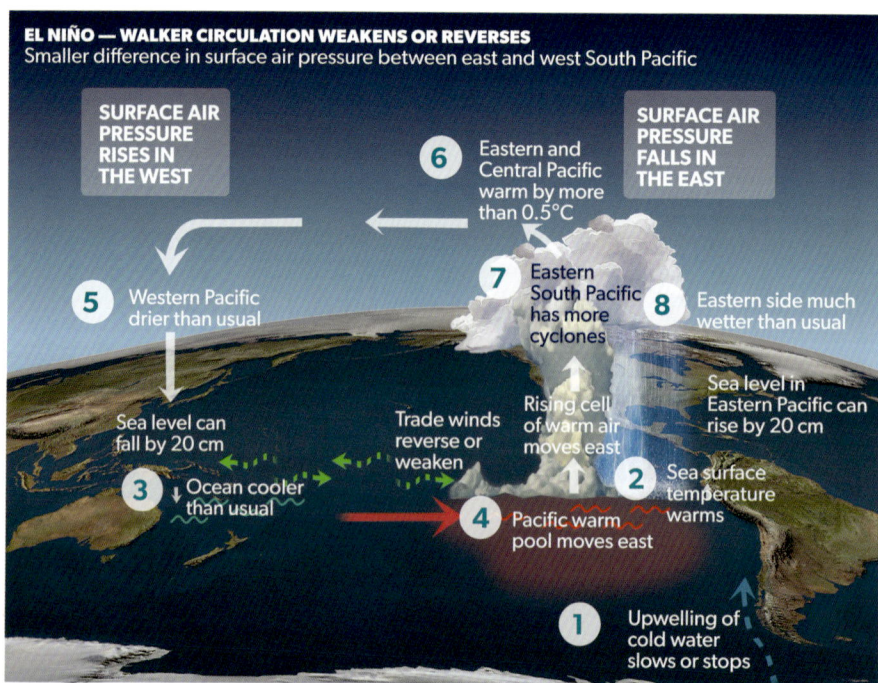

LA NIÑA

La Niña is a general strengthening of the 'neutral' Walker Circulation pattern. It occurs when the difference in surface air pressure between the western and eastern Pacific increases. With stronger trade winds, the western side becomes warmer, the central Pacific is wetter and the eastern Pacific and South America are colder than normal, with less rain. Although La Niña is known as the 'cool' phase of ENSO globally, it brings generally warmer weather to New Zealand and makes marine heat waves like the one in the summer of 2021–22 more likely.

HOW LA NIÑA AFFECTS AOTEAROA WEATHER

The Tasman Sea is warmer than usual and New Zealand has generally higher temperatures • The north can experience more storms • Western and southern areas are drier and can suffer drought • Eastern and coastal areas are often windier and cloudier • A high often sits to the east of the country, bringing more northeast winds than usual in summer and making northern and eastern areas warmer, wetter and more humid.

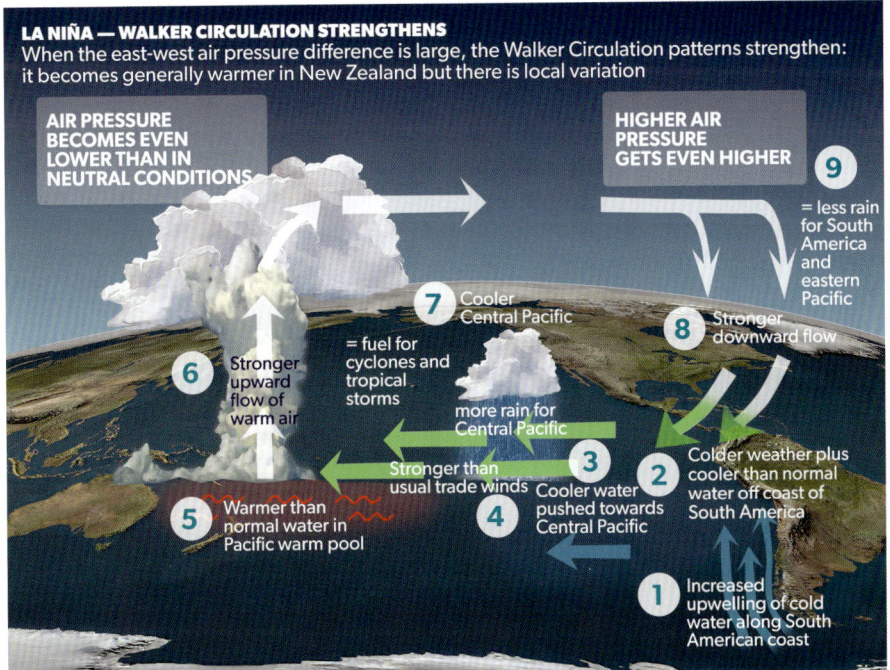

OTHER OSCILLATIONS

There are two other oscillations or phases you might hear meteorologists mention in connection with New Zealand weather. They work over very different time frames to ENSO.

The *Interdecadal Pacific Oscillation* (IPO) lasts 20–30 years and can affect the strength and timing of El Niño and La Niña. It relates to changes in sea surface temperature far away in the North Pacific Ocean. In 2022 it entered a phase that can bring stronger west to south winds and more rain and cloudy weather to western areas. Northern and eastern parts of the North Island can be drier and sunnier.

The *Southern Annular Mode* (SAM), on the other hand, shifts rapidly and unpredictably. It is monitored by measuring changes in sea level air pressure in the band of mid-latitude westerlies in our hemisphere. When the low pressure belt to our south moves closer to the Equator we can get more frequent westerlies; Aotearoa's weather is more unsettled and stormy and the West Coast can be wetter and cooler. When the low pressure belt shifts further south, towards Antarctica, we tend to get fewer and weaker westerlies in early spring.

> **HEAT ALERT** Northwesters can bring record high temperatures to Canterbury. In January 2021 a heat alert was issued, with health warnings on how to avoid overheating in a significant heat event. People were told to stay in the shade, avoid extreme physical exertion and not leave children or pets in parked vehicles. It was recommended to drink plenty of water and leave windows open and blinds and curtains closed for shade. Fire risk is much greater in high temperatures, especially if strong winds are forecast. Even overnight temperatures were in the high twenties. Local heat records were broken: Ashburton reached 39.3°C and Christchurch 37.1°C. The record temperature nationally is 42.4°C in Rangiora, North Canterbury, in 1973.

> **TEMPERATURES** Not only are summer temperature highs getting higher as the world warms, winter temperatures are rising too. In 2021, June was the hottest June in Aotearoa since weather records began. It was a whole two degrees warmer than the norm and 24 places around the country experienced record-breaking highs. Motueka's temperature was 3.2 degrees higher than its June average for 1981–2010. Crops that rely on frost for part of their growing cycle can suffer in warmer than average winters like these. Pests and diseases that are normally killed off by cooler winter temperatures thrive and spread.

SEA BREEZES AND LAND BREEZES

SEA BREEZES
At the other end of the scale there are local wind systems like sea breezes that spring up at the beach on sunny afternoons. Land and sea warm up and cool down very differently. A sea breeze is an onshore breeze that develops once the land has had a chance to heat up more than the neighbouring coastal water.

On calm, still mornings at the coast, as the sun climbs higher in the sky and warms the land, the air above the ground warms too and begins to float upwards. This creates a difference between the air pressure above the land and that over the sea, so colder, heavier, higher-pressure air over the sea comes in to take the place of the rising air, creating a sea breeze.

In a small-scale convection current, the warm air cools as it rises into the sky and, looping back, starts to sink towards the surface of the sea. The larger the temperature difference between land and sea, the stronger the sea breeze. Sea breezes typically start between late morning and mid-afternoon and fade away later in the day as the sun's heat weakens.

The difference in air pressure over land and sea, high above the surface, is called a pressure gradient and it creates the flow of air from the high pressure area to the one of lower pressure.

Sea breeze

SEA BREEZE CONVERGENCE
The weather in Tāmaki Makaurau/Auckland is especially changeable because the city sits on and around an isthmus, a narrow strip of land with sea on both sides.

At its narrowest point the land is less than 2 km wide.

When sea breezes from opposite coasts meet over land, sea breeze convergence — the clash of opposing masses of moist, rising air — can trigger thunderstorms. The same can happen above the narrow Northland peninsula.

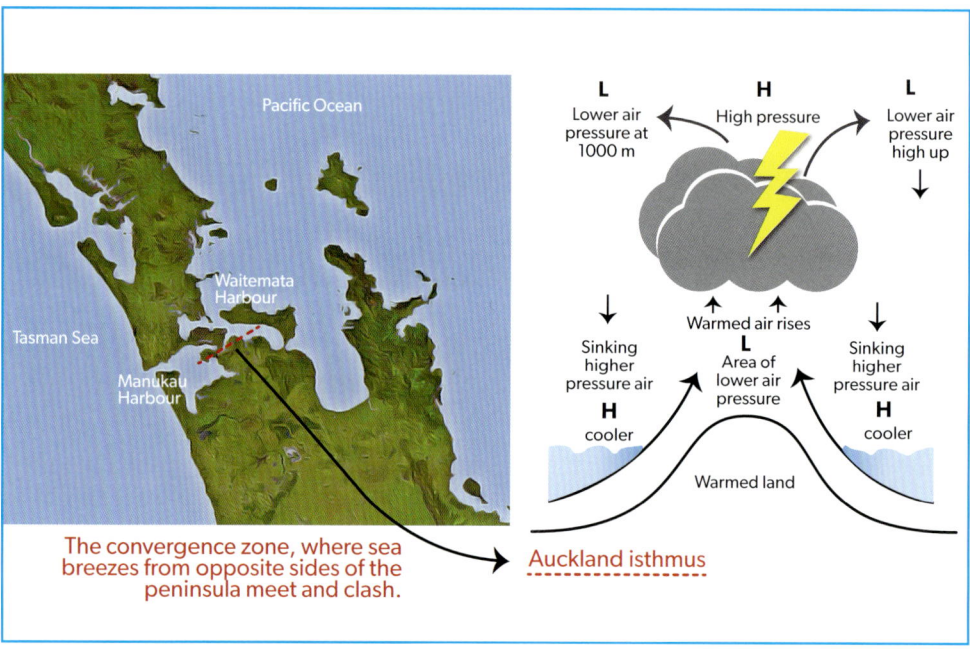

The convergence zone, where sea breezes from opposite sides of the peninsula meet and clash.

Auckland isthmus

Solar energy spreads deeper through water than it can through ground, so the ocean both warms up and cools down more slowly than the land. At night the land's heat radiates back into space and it cools faster than the sea. This gives rise to a land breeze: now the warmer, lighter, lower pressure air over the sea moves upwards and, creating a cell of air moving in the opposite direction, the cooler, heavier air above the land moves in to replace it.

Land breeze

CLOUDS

Clouds are a visible sign of what is happening in the atmosphere. Meteorologists record the amount and type of cloud to help them make forecasts. Clouds are grouped according to their different shapes and the height at which they appear in the sky. The names of the different types of clouds come from Latin words that describe their appearance or height.

WHY ARE RAINCLOUDS DARK? In developing rainclouds, as the number of water droplets grows and the tiny droplets combine to form larger ones, less sunlight can penetrate and reflect and so the clouds darken.

Cirrus means a lock of curly hair; these high, wispy clouds are made of ice crystals. Alto means 'middle'; these clouds are found at mid-height in the troposphere. Cumulus means pile or heap; these are the puffy clouds that float at about 500 metres. Stratus means layer and is a featureless, grey sheet of low cloud that stretches in horizontal layers close to the ground,

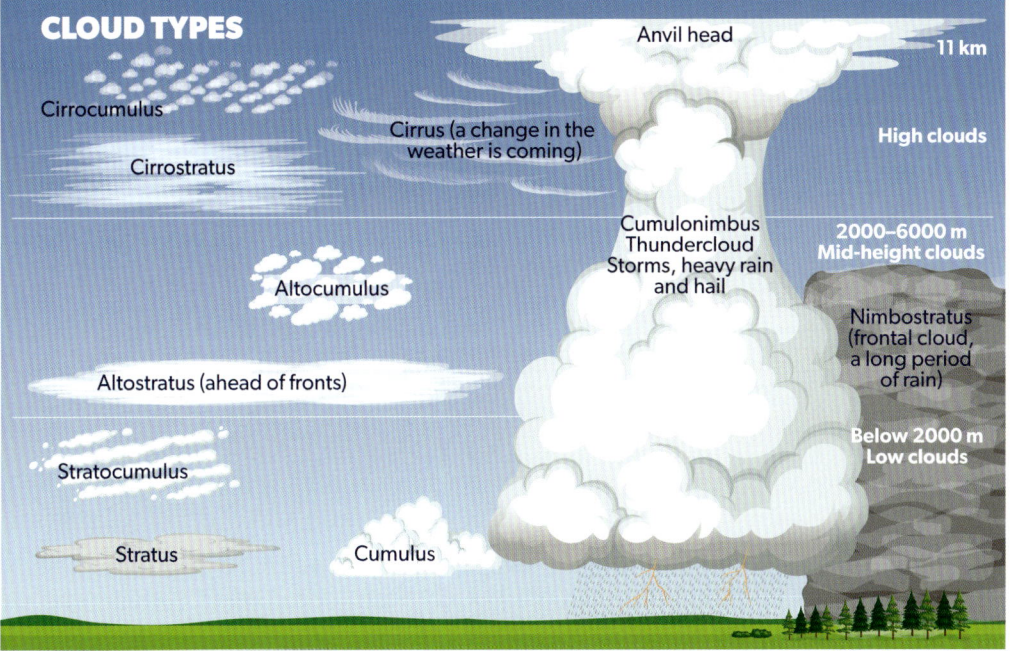

> **AUTOMATIC WEATHER STATIONS** (AWSs) use instruments called cloud ceilometers to record the amount of cloud and the 'ceiling' or height of the cloud base. They work by firing a low-powered laser beam upwards. The beam bounces off the bottom of the cloud layer and back to the ground unit, and that measurement is used to calculate the height of the cloud, both by day and night. They can also measure the concentration of aerosols in the atmosphere and map and track volcanic ash clouds, especially around airports.

often blocking the sun. Nimbostratus is the thick, dark layer of raincloud that stretches for kilometres and can bring days of rain. Cumulonimbus are storm clouds that develop vertically and can reach to the tropopause. Once they reach the stratosphere their top flattens out into the shape of an anvil. Convection cumulus clouds begin to form at the condensation level, which is the height where water vapour in the air condenses.

There are many combinations of basic cloud types, blending their various characteristics (e.g. cirrostratus — high wispy layer cloud) and many sub-branches of each (e.g. altocumulus castellanus, which tower up like castles, and stratus fractus — fast-changing clouds torn by gusty winds). The International Cloud Atlas was first published in 1896 and contains more than one hundred different kinds of clouds.

The thickness of the troposphere varies with both season and latitude, so the height of the different cloud types can vary too. Over southern parts of the South Island the troposphere is around 11 km so ice crystal clouds form at a lower altitude in the South Island than they do in the North Island.

WHY IS THE SKY BLUE?

The air molecules in the atmosphere bend and scatter sunlight. The blue-violet light (which has the shortest wavelength) is bent and scattered the most, which makes the sky look blue by day.

THUNDERSTORMS

When fluffy, white, fair-weather cumulus swells upwards into towering cloud masses and grows dark and heavy-looking, there is a good chance of a thunderstorm coming. Cumulonimbus clouds are the tallest in the sky and the only ones that produce hail, thunder and lightning.

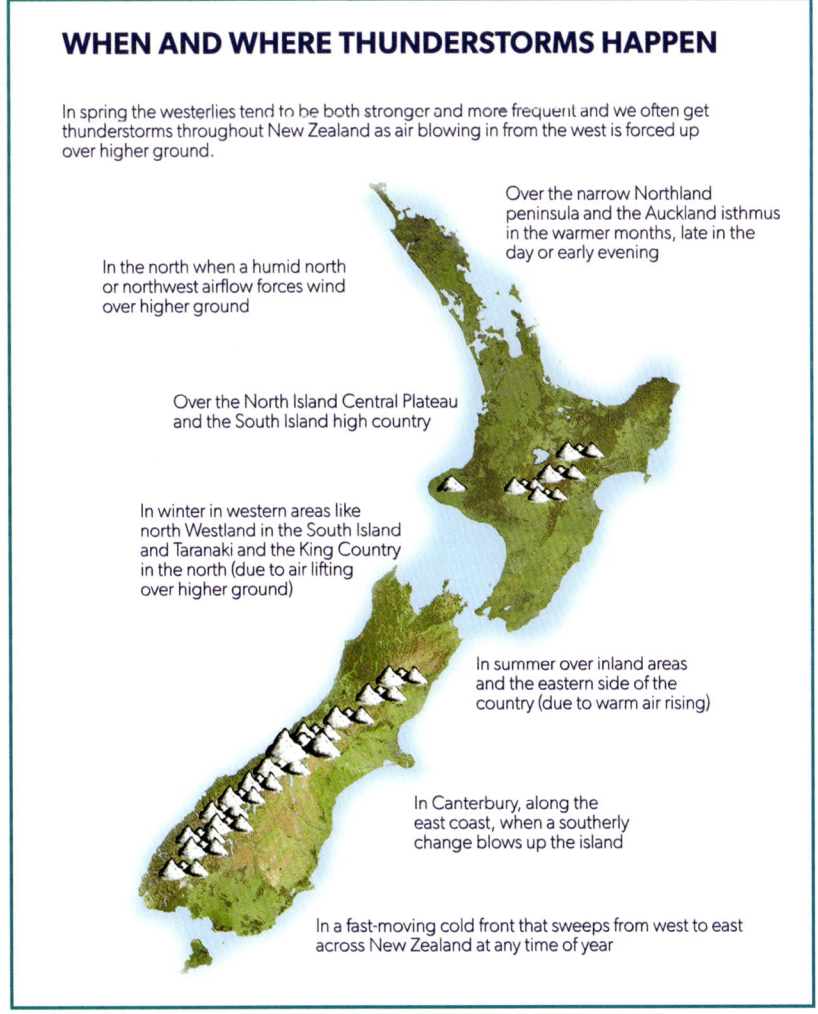

WHEN AND WHERE THUNDERSTORMS HAPPEN

In spring the westerlies tend to be both stronger and more frequent and we often get thunderstorms throughout New Zealand as air blowing in from the west is forced up over higher ground.

- Over the narrow Northland peninsula and the Auckland isthmus in the warmer months, late in the day or early evening
- In the north when a humid north or northwest airflow forces wind over higher ground
- Over the North Island Central Plateau and the South Island high country
- In winter in western areas like north Westland in the South Island and Taranaki and the King Country in the north (due to air lifting over higher ground)
- In summer over inland areas and the eastern side of the country (due to warm air rising)
- In Canterbury, along the east coast, when a southerly change blows up the island
- In a fast-moving cold front that sweeps from west to east across New Zealand at any time of year

> **SIGNS A STORM IS COMING**
> Tall clouds gather • the base of the cloud grows very dark • wind picks up • air pressure drops and the barometer falls • humidity rises and the atmosphere feels hot and sticky • look out for the flat head of the cumulonimbus, shaved by jet stream winds.

Thunderstorms develop when warmer air rises rapidly in an unstable atmosphere; when the air at middle or upper levels of the troposphere is much colder than air near the ground. In the hotter months they can form when opposing sea breezes converge (meet) over land or when air lifts as it passes over higher ground; or, at any time of year, when a fast-moving cold front moves in and forces an area of milder air to rise.

THE STAGES OF A THUNDERSTORM

Thunderstorms usually last for an hour or two at most. They often develop in stages, beginning as puffy, white cumulus. They start out over a small area of the surface, quite close to the ground but they can build to over 10 km high.

The cumulus stage
Fuelled by lots of warm, moist air near the surface, strong updraughts (upward air currents) rise rapidly in the atmosphere. As the rising air cools to its dew point, the moisture it holds condenses into water droplets. This releases heat energy (latent heat) which warms the air in the cloud and provides more fuel to both strengthen and speed up the updraughts, and so the cloud continues to grow upwards.

The mature stage
The cloud becomes dark and heavy-looking as more water vapour condenses and droplets grow and join together to form raindrops. When these are too heavy to float on the rising air, they begin to fall. At the same time, cool, dry air flows down through the cloud, in a downdraught, which pulls the raindrops down. Once the cumulonimbus reaches the tropopause, the top of the cloud spreads out sideways, in the shape of an anvil. Heavy rain falls and powerful downdraughts can cause microbursts — strong winds that blow outwards from the base of the cloud.

The dissipating (decaying) stage

The storm begins to weaken when the downdraughts, which pull the raindrops down through the cloud, become stronger than the updraughts. The heavy rain (or hail or snowfall) cools the ground and the air beneath the cloud; warm, moist air stops rising, cloud droplets stop forming and the rain eases as the anvil head fades into cirrus and the cloud disappears from the bottom up.

> **SEVERE THUNDERSTORMS** are defined as producing one or more of the following:
> - heavy rain of at least 25 millimetres per hour
> - large hail of 20 millimetres diameter or more
> - wind gusts of 110 km/h or more
> - tornadoes with wind speeds of at least 116 km/h. (NIWA)

THUNDER AND LIGHTNING

As the moving currents of air slide past each other in a mature cumulonimbus cloud, ice crystals and frozen water droplets build up electrical charges. Positive and negative charges collect in different parts of the cloud: a negative charge forms at the base, while a positive charge builds near the top of the cloud,

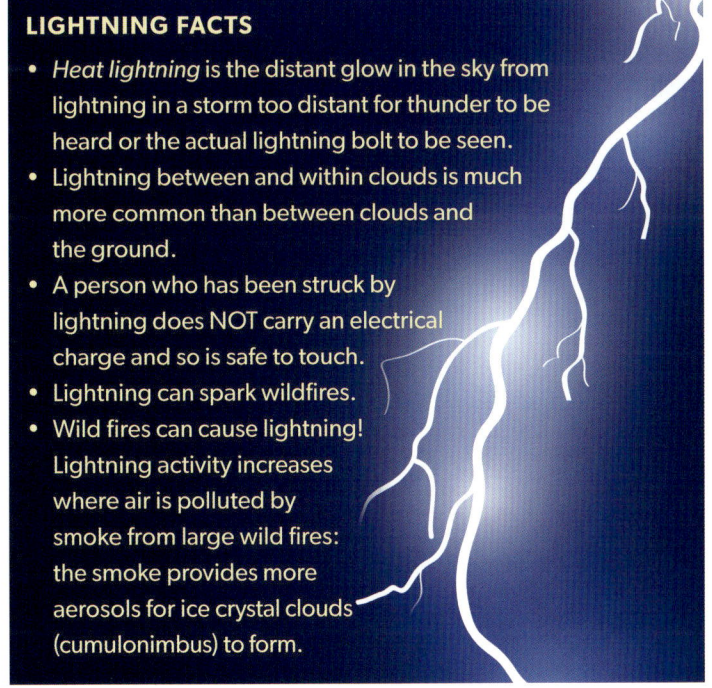

> **LIGHTNING FACTS**
> - *Heat lightning* is the distant glow in the sky from lightning in a storm too distant for thunder to be heard or the actual lightning bolt to be seen.
> - Lightning between and within clouds is much more common than between clouds and the ground.
> - A person who has been struck by lightning does NOT carry an electrical charge and so is safe to touch.
> - Lightning can spark wildfires.
> - Wild fires can cause lightning! Lightning activity increases where air is polluted by smoke from large wild fires: the smoke provides more aerosols for ice crystal clouds (cumulonimbus) to form.

> **LIGHTNING STRIKES** There are an average of 47,000 lightning strikes a year in Aotearoa and more than 100,000 if you include the surrounding seas. Some years, like 2013, there are many more than others. On 30 August 2021 there were over 600 strikes in and around Auckland in just a few hours. Lightning strikes are more common in western parts of the country, especially in the South Island.

where the ice crystals are. The electrical charges build up until a giant spark jumps from one part of the clouds to another, or between the clouds and the ground, and the electricity is discharged as lightning. The differences between the charged regions are temporarily equalised until the electrical charges build up again and the process repeats.

HOW CLOSE IS THE STORM?
Count the seconds (1001, 1002, 1003 …) between the lightning flash and the thunder. Divide the number of seconds by 3 and the answer tells you how many kilometres away the storm is. For example, if you count to 6, the storm will be 2 km away.

Lightning moves at such a speed that it superheats the air immediately around it to about 30,000°C, many times hotter than the surface of the sun! A lightning bolt speeds through a channel of air just 2 to 3 centimetres wide, which, as it heats, expands so fast that it explodes, then immediately cools and contracts, creating the sound waves that we hear as thunder. Thunder and lightning happen at the same time but we see the flash before we hear the thunderclap because sound travels much more slowly than light.

SUPERCELLS AND SQUALL LINES

As one storm cell dies away, another often forms close by. Squall lines are a narrow band of thunderstorm cells that form along or ahead of a cold front. Squalls, on the other hand, are sudden short-lived bursts of high winds with torrential rain.

Supercells, or rotating thunderstorms, have a tall, deep, rotating updraught, called a mesocyclone. Although not unknown in Aotearoa, they are not especially common here.

A mature thunderstorm with its anvil of ice crystal cloud, strong up- and downdraughts and the build-up of positive (+) and negative (-) electrical charges that bring lightning.

HAIL

Hail can form in cumulonimbus cloud when the storm cloud builds to well above the freezing level. The highest parts of the cloud are full of super-cooled water droplets and ice crystals, and raindrops can freeze into tiny balls of ice. Hailstones grow by building up layers. The ice pellets are swept up and down (and sideways) on powerful air currents, colliding with water droplets which freeze on to them.

The layers of ice can be cloudy or clear, depending on where in the cloud and how they form. When liquid water freezes instantly onto a hailstone (a process called riming), air bubbles become trapped in the newly forming ice, making it cloudy. When the water freezes slowly the coat of ice is clear as the bubbles escape. When the updraughts weaken or the hailstones grow too heavy to be held up on the air currents, they fall to the ground.

> **SEVERE HAILSTORMS** can occur anywhere in Aotearoa but especially in western areas and the far south of the South Island. Most of the hailstones are small (the largest in New Zealand are 2–3 centimetres at most) but they can still cause enormous damage to glasshouses, crops and orchards, especially in the spring and summer.

TORNADOES

A tornado is a spiralling funnel of air that breaks free of a storm cloud and whirls wildly over the surface, sucking up and destroying everything in its path. New Zealand tornadoes are nowhere near as frequent or powerful as those in the United States, but they can happen all year round. In the last few decades several people have been killed and many more injured by tornadoes in Aotearoa.

HOW A TORNADO DEVELOPS

Tornadoes can happen when two different winds moving at different speeds and levels above the ground meet in the same place. This is called *wind shear*. It can occur when there is moist air near the surface and cold, dry air above. An increase in wind speed or a sudden change of wind direction can set the air rotating in a horizontal, twisting cylinder that gets tilted upright when warm air rises.

Inside a storm cloud this spiralling tube develops into a funnel cloud, which extends from the base of the cumulonimbus and stretches down to the ground. Once it touches the surface and breaks free it becomes a tornado. Recent research suggests that tornadoes originate not only within a storm cloud but can develop from the ground up.

> **TORNADOES IN AOTEAROA**
>
> On average around 10–20 tornadoes and waterspouts are reported to MetService each year, but most of these are small. They occur mostly in the cooler months, most commonly in western areas such as Westland, Taranaki, Auckland and Northland, and in Waikato and the Bay of Plenty.

TORNADO WARNINGS

Weather radar is used to track thunderstorms but is unable to pick up small-scale phenomena such as tornadoes. Meteorologists can forecast the likelihood of severe thunderstorms and issue thunderstorm outlooks, watches and warnings and say if there is a tornado risk. If a thunderstorm is less than 150 km from weather radar, MetService meteorologists can confirm whether it can be classed as 'severe'. A thunderstorm does not have to be classed as severe for it to spawn a tornado.

The tornado that struck Papatoetoe, South Auckland, in June 2021 left one person dead and more than 200 homes damaged.

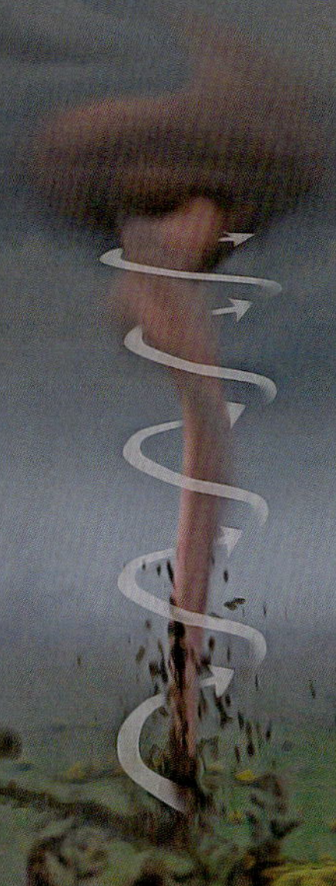

TORNADO FACTS

- The high-speed winds in a tornado swirl around a centre of extremely low pressure.
- They rotate clockwise in our hemisphere and anti-clockwise in the north.
- A waterspout is the equivalent of a tornado over water.
- Sometimes a narrow band of thunderstorms (a squall line) ahead of a cold front can give rise to multiple tornadoes. In July 2007 a dozen or more tornadoes struck the Taranaki coast, causing widespread damage.
- Tornadoes are usually between 20 and 100 metres wide and cover a distance of around 2–5 km. The one that struck Albany, Auckland, in May 2011 had a 15-km track and caused $10 million worth of damage.

HOW WEATHER SYSTEMS FORM

Air pressure is not the same all over Earth's surface. The circulating cells of rising and sinking air that transport the Sun's heat around the globe create loose bands of higher and lower pressure air at different latitudes.

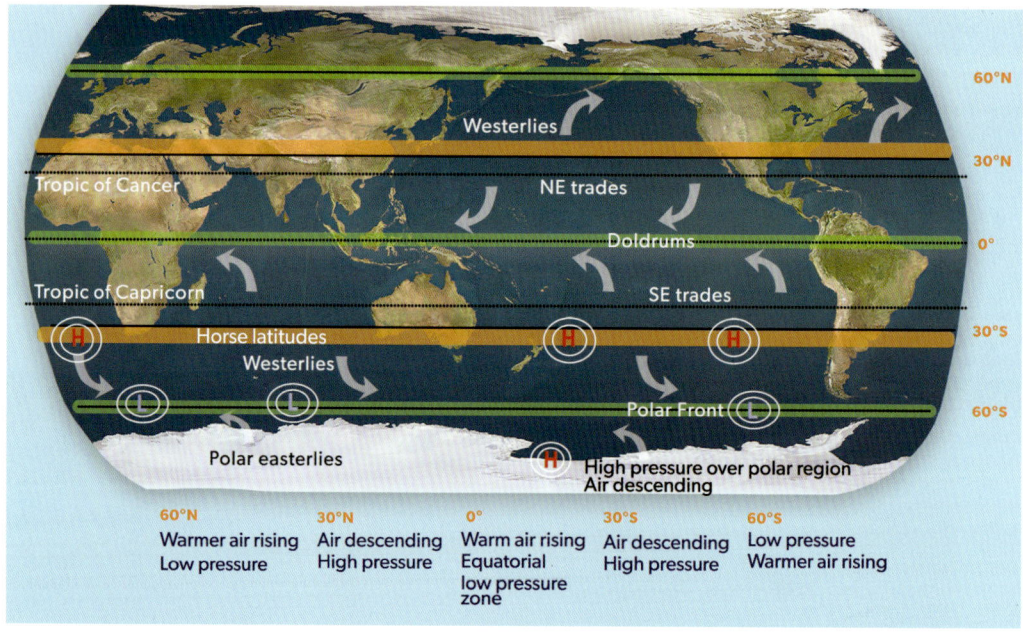

MASSES OF AIR

An air mass is a large body of air that has broadly the same temperature, density and humidity across every level of altitude. Air masses form in still conditions, mainly in the high pressure belt in the subtropics. Over many days, the characteristics of the air at the surface, whether ocean or land, slowly spread throughout the mass. Air masses globally can be tropical or polar, maritime or continental. Surrounded by ocean, rather than a continental landmass, New Zealand experiences just two different types of air masses: tropical maritime ones mainly from the north and the northeast, and polar maritime ones from the south and southwest.

Air masses can stretch horizontally for hundreds, even thousands of kilometres, as well as vertically through the entire height of the troposphere. The movement of air masses around the globe is driven by changes in the jet stream winds in the upper atmosphere, which push them around. When an air mass moves away from the area where it formed its temperature, humidity and stability change.

> The *horse latitudes* is a traditional sailors' name for this region of fine, calm weather and light winds, where ocean currents would speed a sailing ship along like a galloping horse. In contrast, the *Doldrums*, around the Equator, is a windless area where sailors are often becalmed.

HOW HIGHS FORM

New Zealand weather tends to alternate between fine spells, when an anticyclone or High moves in from the tropics, and wind, rain and lower temperatures as a front or a Low passes. A High is an area of higher than average air pressure. High-pressure systems often form in the horse latitudes, to the north of Aotearoa — the high pressure zone where the air in the Hadley and Ferrel cells sinks backs towards the surface. The warm air moving out from the Equator rises high into the troposphere, losing moisture and becoming cooler as it does. Highs begin to form as powerful jet stream winds push air around in the upper atmosphere. As the cool, dry, heavy air in the centre of a high pressure area sinks towards the ground, the Coriolis effect pushes it into a spiral.

Air compresses (the molecules get squeezed closer together) as it sinks and the compression warms the air, like air pumped into a bicycle tyre. As the air warms, the water vapour it holds evaporates, so clouds disappear and the sky clears.

Air flows naturally from areas of high pressure to ones of lower pressure and vice versa, evening things out, so near the surface, air flows outwards from an area of high pressure, towards lower pressure. The Coriolis effect deflects this air so, in our hemisphere, the winds circulate anticlockwise around a High.

GLOOMY HIGHS

Highs usually make you think of fine, dry, settled weather and clear skies but they can also bring anticyclonic gloom. The descending air in a High does not usually sink all the way to the surface but spreads out at about 1000–1500 metres above sea level. Sometimes this warm air can trap a layer of cold air near the surface, creating an inversion that prevents convection. The air that is rising just

> **SUNSHINE HOURS: RECORD HIGHS AND LOWS**
> Richmond   2840 hours in 2016
> Whakatāne   2792 hours in 2013
> Invercargill   1333 hours in 1983
> Palmerston North 1357 hours in 1992
>
> Invercargill had only 35 hours of sunshine in June 1935
>
> Nelson and Marlborough enjoy amongst the highest sunshine hours in the country, often more than 2500 annually. These regions are sheltered from the wet westerlies and cold southerlies by hills and mountains to the west and south. Whakatāne, Tauranga, Gisborne, Napier and Central Otago also have high sunshine hours but in 2021 New Plymouth topped the list in NIWA's annual climate report, clocking up 2592 hours over the year.

above the surface stops rising any further because it is less warm than the upper layer. Water vapour and aerosols trapped in the cooler layer create a blanket of cloud, which, unable to spread upwards, is forced to spread outwards. If the wind does not pick up and the sunlight is too weak to evaporate the cloud, the stratus and stratocumulus can linger all day.

BLOCKING HIGHS

High-pressure systems can also remain stationary over an area, stopping other weather systems from moving in and keeping the weather the same for days or even weeks on end. Often the blocking High sits to the east of New Zealand, holding up the usual irregular succession of lows, troughs and ridges that come from the south and west, and causing heatwave conditions.

HOW TO READ A WEATHER MAP

WEATHER CHARTS: ISOBARS AND HIGHS

Synoptic weather charts illustrate the weather at a given moment in time, showing the positions of weather systems like Highs, Lows and fronts on a map. The lines on weather maps are isobars. They join places of equal surface air pressure. When an isobar is labelled with a number, then that is the pressure in hectopascals (hPa) along the whole length of that line. An isobar to one side of the line will be lower pressure, and to the other side it will be higher. In New Zealand the spacing between isobars is usually 4 hPa, but not every line is numbered.

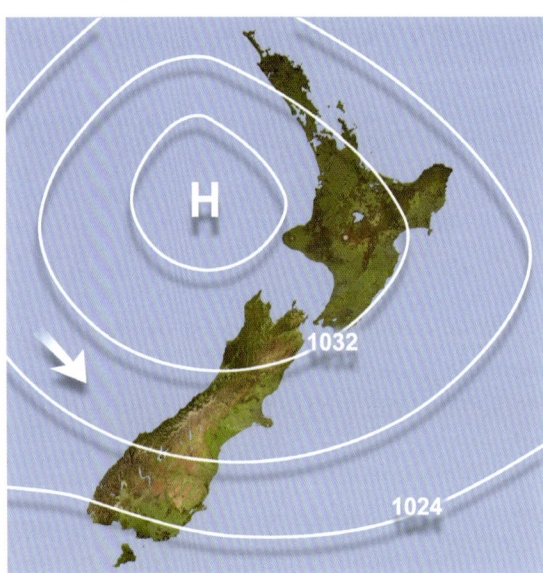

An anticyclone or high is shown as a ring of isobars with an 'H' in the middle. The isobars are widely spaced which means that winds are generally light.

If you picture a column of air pressing down onto the surface, the column of air above a West Coast beach would be taller than the one above the Southern Alps. The air pressure readings on weather maps are corrected to sea level so that the measurements are on an equal footing and take the contours of the land into account.

A ridge of high pressure

A ridge is an elongated area of high pressure that usually means warm, dry weather. There is no symbol for a ridge on a weather chart. When the isobar at the centre of a High looks more like a stretched oval than a circle, the tongue-shaped area extending from it is a ridge. A High can have several ridges and a ridge of high pressure can also divide two low-pressure regions.

LOWS

The depressions or Lows that affect our weather often form in the belt of rising air to the south of us, where warmer air meets cold polar air. Lows move much faster than Highs and the Deep South often has more unsettled and fast-changing weather than the north of the country. Southland and coastal Otago lie in the path of the chilly, windy weather systems that blow in from the south and southwest. They tend to have more cloud and lower sunshine hours than other regions, typically as low as 1600 hours a year.

On the Southland coast trees are shaped by the strong winds that blow in from across the ocean.

How a Low-pressure zone forms

A Low forms as different air masses push around each other and make a swirl of rising air. The air spirals inwards and then upwards. The rising air leaves an area of low pressure. Surface winds blow clockwise into the Low, replacing the rising air.

As the warmer air in the centre of the Low rises, it cools; the water vapour it holds condenses and forms clouds. There are often tall cumulonimbus clouds around a Low, as well as altocumulus, altostratus, cirrocumulus, stratocumulus and stratus clouds.

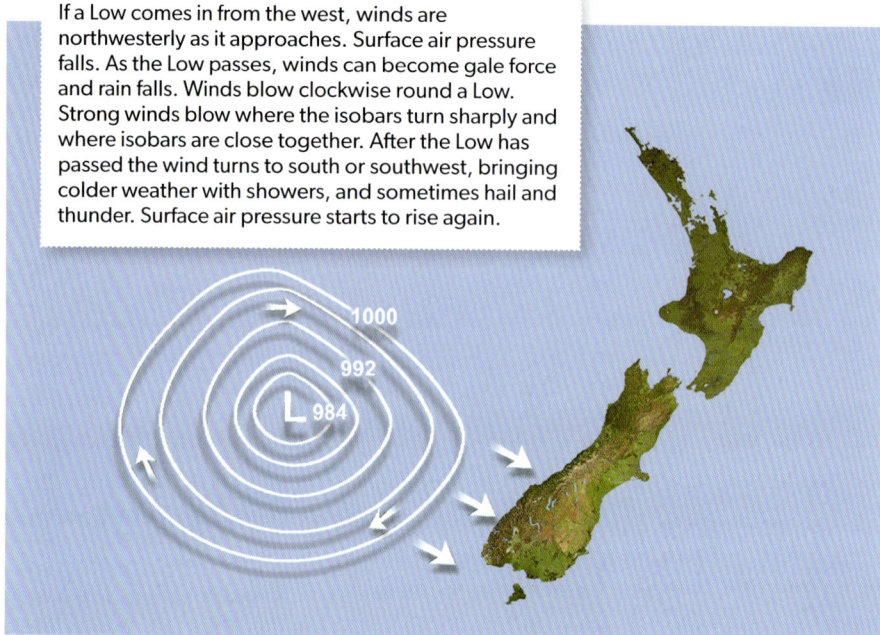

If a Low comes in from the west, winds are northwesterly as it approaches. Surface air pressure falls. As the Low passes, winds can become gale force and rain falls. Winds blow clockwise round a Low. Strong winds blow where the isobars turn sharply and where isobars are close together. After the Low has passed the wind turns to south or southwest, bringing colder weather with showers, and sometimes hail and thunder. Surface air pressure starts to rise again.

Isobars are just a guide to wind direction and strength. A depression is an area of lower-than-average air pressure and is labelled with an 'L' on weather maps. A complex Low is one with two or more centres. When the air pressure in the centre falls, the Low is deepening or intensifying. When the central air pressure rises, the Low is described as weakening, or filling. A Low brings cloud, rain, wind and unsettled weather, can spread over hundreds of kilometres and take a day or two to pass.

TROUGHS

A trough is an elongated area of relatively low pressure that extends away from a Low and brings the same kind of cloudy, wet weather. A Low can have more than one trough associated with it. Troughs can occur in the jet stream, in the upper levels of the atmosphere, or close to the surface, where they usually accompany fronts.

FRONTS

Fronts form when two air masses of contrasting temperatures push up against each other, forming a boundary line.

HOW TO READ A WEATHER MAP

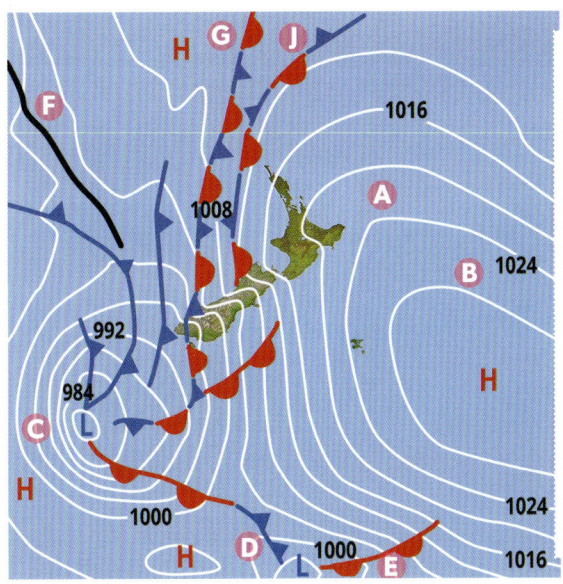

A a ridge of high pressure stretches out northwestwards from the centre of the High
B widely-spaced isobars around the High mean light winds
C strong winds where isobars are close together
D cold front — cold fronts often accompany Lows
E warm front
F trough of low pressure
G **Occluded front** — a cold front overtakes a warm front, bringing increasing cloud and possible rain
J **Stationary front** — weather changes very slowly, rain clears very slowly

Fronts and Lows are often linked on weather maps. Fronts can be 2–3 km thick and stretch for hundreds of kilometres. They often move from west to east across Aotearoa and can be the centres of large areas of low pressure.

Cold fronts

There are different kinds of weather fronts. Most fronts that move across Aotearoa are cold fronts. They form when a cold maritime polar air mass catches up with a slower-moving maritime tropical air mass. Cold fronts can move up to twice as fast as warm ones. They usually come from the southwest and move northeastwards across the country. The steeper the slope of a cold front, the more active it is. Fronts often take only 2 to 3 hours to pass, bringing heavy rain and sometimes thunderstorms too.

Warm fronts

Warm fronts in our hemisphere are much shorter in length than those in the northern hemisphere and

SYMBOLS for cold **D** and warm **E** fronts point in the direction the front is travelling. In an occluded front **G** the triangles and semicircles point in the direction of travel and are on the same side of the line. In a stationary front **J** the triangles and semicircles alternate along opposite sides of the line, with the triangles pointing into the warmer air mass and the semicircles facing into the cooler air mass.

THE STAGES OF A COLD FRONT

If you could cut a slice down through the atmosphere, you would see the cold front as a sloping wedge of cold, heavy air pushing a wedge of lighter, warm air upwards and replacing the warm air at the surface.

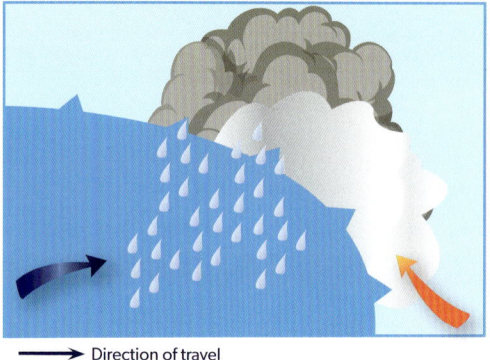

Direction of travel

1. Warm northwesterly winds blow and high cirrus clouds can appear ahead of a cold front.
2. The temperature drops and low stratus and stratocumulus form, making the sky overcast. It can start to drizzle.
3. As the heavier, denser, cold air pushes under less dense, warm air, taller cumulus-type clouds develop, sometimes with cumulonimbus.
4. Winds become gusty and the temperature suddenly falls. Heavy rain, possible thunderstorms and hail follow.
5. The wind shifts sharply to a strong, cold sou'wester. This is called a southerly change. The skies clear and a few days of cold, bright weather often follow once the front has passed.

A CROSS SECTION OF A WARM FRONT

Direction of travel

1. The weather ahead of a warm front is cool and cloudy, often with increasing showers and rain.
2. The bank of cloud that accompanies a warm front slopes forward from the ground upwards. The slope is far more gradual than the angle of a cold front and a warm front moves more slowly.
3. The clouds are mainly layer clouds of the stratus type, which develop from the slow, gradual lifting of air. Stratus, cumulostratus, altostratus and nimbostratus start to thicken from up to 12 hours ahead of the front, becoming increasingly low in the sky.
4. After the front passes the temperature can rise a little or become steady. The wind shifts to the northwest with overcast conditions, high humidity and light rain or drizzle for 12 to 24 hours before skies eventually clear.

are also called 'warm sectors'. They can come from the northwest and move southeastwards over New Zealand.

Occluded front

An occluded front (or occlusion) is formed when a cold front catches up with a warm front and the warm air is forced up from the surface. As it rises and cools it creates a band of thick cloud and rain and the wind shifts. As an occluded front passes the rain and wind ease, skies clear and the air temperature remains much the same.

Stationary front

A stationary front can be a weakening front or form when two different air masses meet and neither one is strong enough to push the other. They move extremely slowly. There is often cloud and rain along a stationary front. The weather can get stuck and lead to days of gloom.

> **EXTREME WEATHER** In Canterbury a warm nor'wester is often followed by a cold front and when the wind changes to a southerly the temperature can drop as much as 20°C in a few hours. There's a traditional Māori saying from the south: 'Kai te taki wahanui ki te toka; kai te tono atu ki te toka to taki', meaning 'the nor'wester calls up the southerly wind'.

Cape Reinga in the Far North.

Cardrona, Central Otago.

WIND

Winds blow stronger in the Southern Ocean than at the same latitude in the northern hemisphere, where land masses slow down the wind. Friction with Earth's surface not only slows wind speed but churns up the air, even though we cannot see it. The ocean surface causes little friction: waves are usually too small to have an effect and often move with the wind. When the prevailing west wind reaches Aotearoa it is forced to rise over obstacles like the Southern Alps, Mount Taranaki and the Tararua and Ruahine ranges, or flow around them.

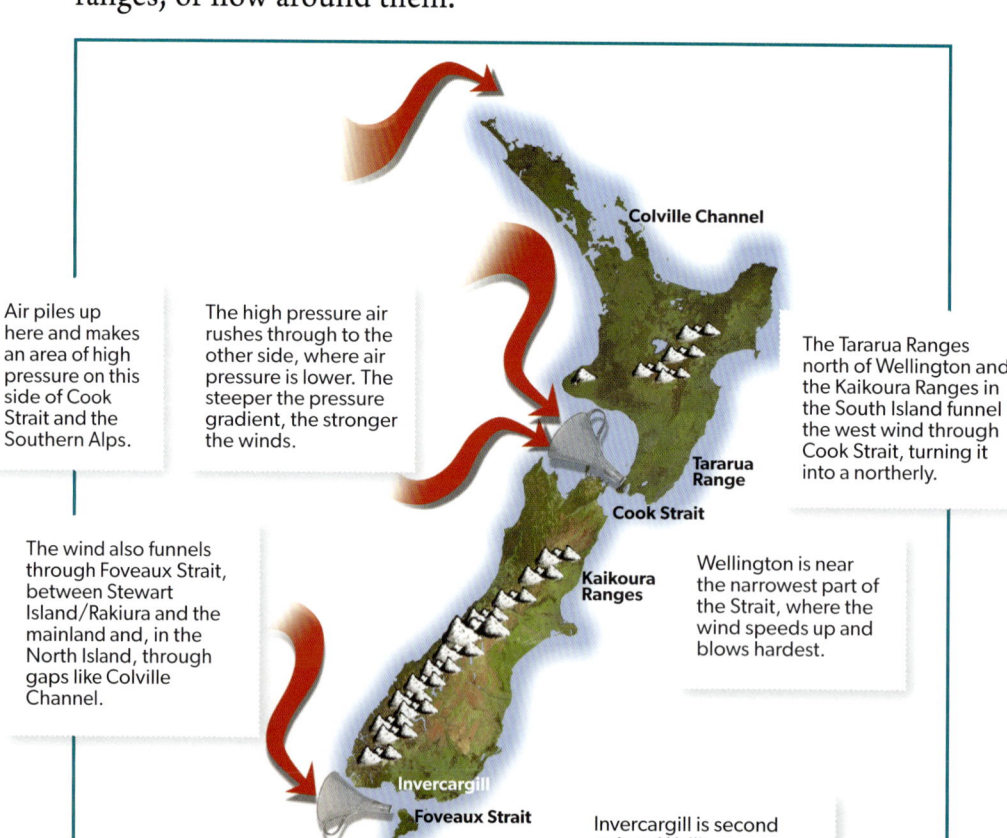

Air piles up here and makes an area of high pressure on this side of Cook Strait and the Southern Alps.

The high pressure air rushes through to the other side, where air pressure is lower. The steeper the pressure gradient, the stronger the winds.

The Tararua Ranges north of Wellington and the Kaikoura Ranges in the South Island funnel the west wind through Cook Strait, turning it into a northerly.

The wind also funnels through Foveaux Strait, between Stewart Island/Rakiura and the mainland and, in the North Island, through gaps like Colville Channel.

Wellington is near the narrowest part of the Strait, where the wind speeds up and blows hardest.

Invercargill is second only to Wellington in terms of strength and frequency of wind gusts.

Wind picks up speed as it crosses the open ocean. Wellington, New Plymouth, Whanganui and Palmerston North are exposed to the westerly winds that blow across the Tasman Sea and, together with Invercargill, are among New Zealand's windiest cities. Inland places like Hamilton, Alexandra and Queenstown are often less windy than coastal areas. A 30 km/h west wind coming in across the Tasman Sea can drop away to 10 km/h at the top of the West Coast but speed up as it rounds Farewell Spit until it is double the original speed as it blows, as a northerly, through Cook Strait.

The wind speeds up even more as it blows through the mountain passes of the Southern Alps and also accelerates through the Manawatu Gorge, which divides the Ruahine and Tararua Ranges. The same 30 km/h wind accelerates around the ends of the Puketoi Range and can end up as a much stronger northwester on the Wairarapa coast near Wellington.

Wind speeds up when it flows downhill or rushes around obstacles or through narrow gaps like Cook Strait. When the westerlies reach the Southern Alps some of the wind lifts and rises over the mountain ranges, some races round the ends of the islands, while some funnels through the gaps between them. The shape of the land can alter the wind's speed and direction and can speed it up or slow it down. Land creates drag; just as a bumpy surface slows a bike or skateboard, so inland regions are often less windy than coastal regions. Wind speed can pick up over flat plains. Winds swirl and eddy over and around towns and higher ground.

WIND SPEED Winds blow at different speeds at different levels of the troposphere. The wind speed given in forecasts is the average over a ten-minute period. A gust is a short burst of markedly stronger wind, usually lasting 20 seconds or less. The New Zealand record is 250 km/h at Mount John in Canterbury on 18 April 1970. The record gust for the Wellington area is 248 km/h in both 1959 and 1962 at Hawkins Hill.
(MetService)

> **ANABATIC WIND**
> Mountain winds can even blow uphill. When a mountainside is warmed by the Sun and the air directly above it heats up more than the surrounding air, an upslope anabatic wind can develop.

WICKED WINDS AND FLYING SAUCERS

Mountain passes are among the windiest places in Aotearoa. Sometimes a layer of stable air at or just below the level of mountain tops prevents air from simply rising over the mountains, forcing it round land obstacles and through gaps, which greatly speeds up the wind. Mountain areas often experience much stronger wind than the weather map suggests; MetService provides separate mountain forecasts.

The west wind bends as it crosses the Southern Alps and changes to a northwesterly as it blows across Canterbury. This kind of strong, dry, hot and gusty wind that blows down the sheltered side of mountain ranges, is known as a foehn (or Föhn) wind, named after a wind in the European Alps. Gisborne, Napier and Masterton, all in the lee of the central North Island mountains, can also experience foehn winds in summer. The foehn dries the crops and soil and is known for making people and animals hot and grumpy in regions already prone to drought. The foehn wind is expected to become more frequent as our climate heats up.

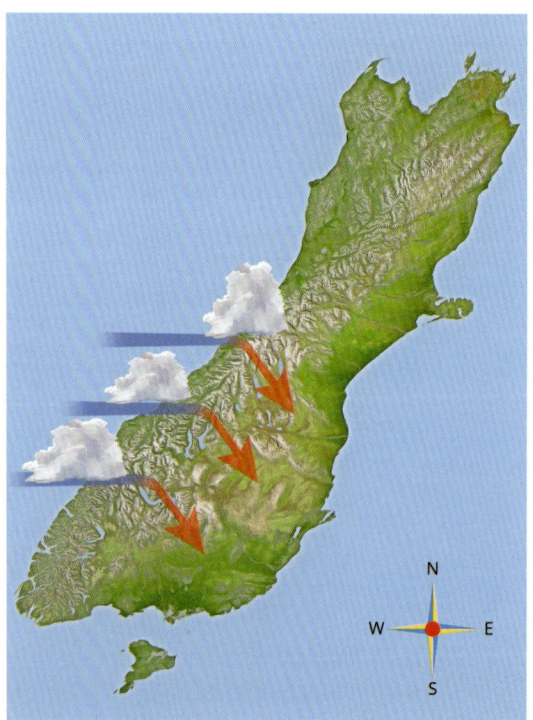

FLYING SAUCER CLOUDS

Lenticular wave clouds can look like flying saucers or stacks of pancakes. They often form over the Tararua Ranges between Wellington and the Wairarapa, and above the Canterbury Plains.

LENTICULAR CLOUDS

The air gets churned up as it crosses the mountains and forms invisible waves that move in up and down patterns. Wave clouds or lenticular (lens-shaped) clouds form on the lee (downwind, protected) side of a mountain range.

Lenticular clouds form in the warm, dry air that has shed the rain it was carrying.

The upper part of the wave is colder than the lower part so water vapour condenses and cloud forms in the air flowing there. As the air moves to the lower part of the wave, the water vapour evaporates and the cloud there disappears.

Even in a strong wind, lenticular clouds seem to hover like flying saucers, but, in reality, the cloud keeps evaporating and a new one reforming in the same place.

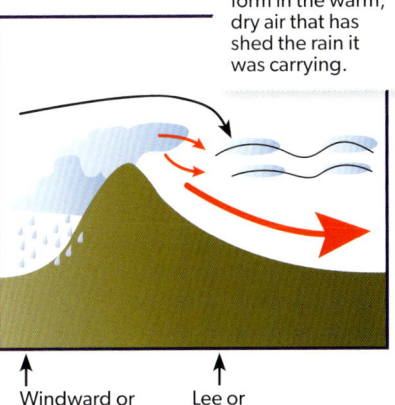

Westerly winds blow in from across the Tasman Sea

Windward or upwind side of mountain range

Lee or downwind side of mountains

THERMALS

Thermals are pockets of warm rising air. The Sun's uneven heating of air near the surface creates these shifting convection currents. Glider pilots and birds such as hawks ride on these updraughts. Thermals can be a bubble of warm air floating free above the ground or an invisible column of air spiralling up from the surface. Dust devils are a visible kind of thermal. Omarama in the Mackenzie Basin in Canterbury is renowned for its thermals and international gliding events. Thermals lose their strength later in the day as the Sun gets lower in the sky.

EXTREME WIND SPEEDS and extreme wave heights have increased globally over the past 30 years. Satellite data, cross-checked with measurements from ocean buoys, shows that the biggest increases have been in the Southern Ocean. Climate warming and changes to the ozone hole are likely to be the cause. Wave heights are 30 centimetres taller and wind speeds 1.5 metres per second faster.

Lenticular clouds.

Katabatic wind

'Katabatic' comes from the Greek word 'katabatikos', which means 'moving downhill'.

At night, cool mountain air flows down valleys and gorges, picking up speed as it races downhill. The steeper the mountain slopes and the bigger the mountain basin, the stronger the wind. The well-known Greymouth 'Barber' on the West Coast (said to be 'so cold it shaves the hair off your face') starts off in the mountains behind the town and can be gale force by the time it reaches the coast and blows out to sea.

Antarctica is known for its biting katabatic winds but in Aotearoa they can arise anywhere that hill or mountain slopes are steep enough for air to become sufficiently chilled. They are most common in winter, on clear, dry nights with snow on the ground.

WIND CHILL Mountain weather forecasts give the 'wind chill temperature' — an indication of what the temperature feels like given the actual temperature and expected wind speed. As a general guide, the temperature can be around two degrees colder for every 10 km/h of wind. So in a 50 km/h wind, it could feel ten degrees lower than the actual temperature.

RAIN

For its size, New Zealand has lots of rivers, mountains and lakes compared to many countries in the world. It has, relatively speaking, ample regular rainfall for growing food, although the rain does not always fall where and when we want. And as the warming atmosphere is able to hold increasing amounts of water vapour, heavy rain events are increasingly delivering more rain at a time than we can cope with. As our climate changes, the greatest change is expected to be in the intensity of extreme weather events, like the torrential downpours and flooding many parts of the country are already seeing.

Weather patterns are likely to stay generally the same, though not entirely, and to intensify. Compared to the South Island, rainfall is more evenly spread in the North Island, but eastern areas, in the lee of high country, are generally drier. The east-west rainfall contrast is far greater in Te Waipounamu/the South Island, because the mountain ranges there are so much taller and longer. The more drought-prone regions, like the Far North and eastern parts of both islands, are likely to see more extended dry spells and the west coast of the South Island is likely to become even wetter as our climate changes.

> **VIRGA** is when falling rain or snow evaporates before it reaches the ground. It consists of streaks of precipitation (rain or snow) that hang from the base of the cloud.
>
>

IS IT GOING TO RAIN?

Whenever air rises, clouds can form and that can lead to rain. Rain falls along weather fronts and around low-pressure systems; it can

THE COST OF CLIMATE CHANGE

Taking 2021 as an example, about 60 severe weather warnings were issued in that year. In 2019, a new Red Warning category was introduced for the 'most significant events' and three were issued by MetService in 2021. There were more high temperature records set than expected, even taking climate change into account.

NIWA notes that the odds of having record-breaking weather events are increasing.

August — Kumeū
Storm — flooding. About 200 mm of rain in 24 hours (but Auckland Airport had less than 20 mm during this period, so it was very localised).

December — Te Araroa and East Cape Tairāwhiti
Flooding, roads closed by potholes and debris.

November — Gisborne
Three times normal rainfall for November in 48 hours plus another month's worth of rain in the following three days. State of emergency declared. Flooding, slips, landslides, mudslides, silt damage.

June — South Auckland
Tornado. Over 1200 homes affected and 60 of them left uninhabitable.

July — Weather event that caused flooding on West Coast brought heavy wind to parts of both islands. Roads left closed in Marlborough at the end of the year.

December — Paraparaumu
Record wet December. 300 mm of rain by 17 December (average for month usually about 70 mm).

July — Westport and Buller
Second Red Warning of 2021. Flooding, 200–500 mm of rain in a few days. Five hundred homes left uninhabitable or in need of repair.

January
Canterbury Health issued warnings about overheating and record high temperatures in Ashburton and Christchurch.

August — Storm
Roads and schools closed. Heavy snow in South Island and even light, fleeting snow flurry in Wellington.

May — Canterbury floods
28 May: MetService Red Warning issued. State of emergency declared 30 May to 10 June. A third of the annual rainfall fell. Extreme flooding.

January— Central Otago, Alexandra Airport
Twice January's normal rain fell in one day. Flooding in Waitaki, Otago and Southland. Roads and bridges closed. Rivers burst their banks.

September — Canterbury
Third Red Warning of 2021. Strong winds. Gale force winds cut power, caused fires and car crashes. Lightning strikes sparked fires in East Otago and Wanaka.

December — Canterbury
140 mm of rain forecast over 21 hours, three times the amount for the whole month.

In flooding, drinking water as well as flood waters become contaminated with waste water when systems are overloaded and unable to cope.

The extreme weather events of 2021 alone lost insurers and insurance policy holders over $300 million.

The cost of not tackling climate change rises every year. The damage caused by the May Canterbury floods was estimated to be over $19 million.

(Compiled from NIWA and MetService data)

fall when updraughts develop over warmed surfaces and, especially, when air is forced to rise over obstacles like hills and mountains.

RAIN SHADOW

The topography (shapes and features of the land) of different parts of the country affects where the rain falls. The Southern Alps are the highest and longest mountain range in Aotearoa so the west coast of the South Island has the highest rainfall in the country. Milford Sound can routinely get over 7000 millimetres (that's 7 metres!) of rain a year, and a total of 16,617 millimetres was recorded for 1998 at Cropp River in the Hokitika catchment.

A fraction of those figures falls on the eastern side of the mountain divide. Alexandra in Central Otago averages around 350 millimetres annually (and in 1964 had a mere 212 millimetres),

1 The Southern Alps run like a backbone along the west coast of the South Island, forming a barrier to the west winds that blow across the Tasman Sea. This relatively warm, moisture-laden air is forced up over the mountains. This is called orographic lifting (orographic = to do with mountains).

2 The rising air expands and cools and air pressure falls. When the air temperature falls to its dew point the water vapour condenses into tiny droplets and cloud forms.

3 When the droplets get too big and heavy to stay in the cloud they fall as rain, sleet or snow.

4 As the air sinks towards the plains to the east, it compresses and warms, like air squeezed into a bicycle tyre.

5 As the air warms, water vapour evaporates from it and it dries. This makes a rain shadow on the lee (sheltered) side of the mountains, bringing fine, dry weather.

while Christchurch has barely over 620 in an average year. In the North Island, New Plymouth, in the west, can average 1400 millimetres, while Napier, sheltered from the westerlies, in the east, has less than 800 millimetres. The westerly winds are generally strongest and most frequent in the spring, so that is when the West Coast can be at its wettest.

HOW RAIN FORMS

Water vapour needs a cold surface to condense onto, like a window or a mirror. Inside clouds fine droplets of water vapour condense or seed onto tiny floating specks called aerosols. These particulates act as condensation nuclei.

Since the 1940s, with varying degrees of success, people in different parts of the world have experimented with artificial seeding of clouds to produce rain, by scattering or shooting chemicals like silver iodide into clouds. The weather touches every aspect of our lives and short-term interventions like these do not seem to be the answer.

Raindrops are not actually raindrop-shaped. When the tiny

droplets condense onto aerosols in the cloud they range from 0.0001 to 0.005 centimetres in diameter. A raindrop, from around 0.5 millimetres across, starts out as round. When it reaches 2 millimetres in diameter it starts to sag and flatten at the bottom. As it gets bigger it sags more and more around the edges and bends in the middle, until, at around 4.5 millimetres in size, it splits into two separate raindrops. And that is the moment they actually look like raindrop shapes.

SNOW

Many parts of Aotearoa never see snow but the rain that falls here often starts out as ice crystals in the clouds and melts on the way down. If the air temperature beneath the cloud is below zero, precipitation falls as snow. When the surface air temperature is just above zero, snow can still reach the ground.

Lake Wanaka.

HOW SNOWFLAKES GROW

High in the coldest parts of the cloud, water vapour freezes into ice crystals. This is called deposition. Ice crystals form different shapes — columns, stars, plates, needles and hexagonal prisms, depending on the temperature and humidity of their environment.

Like raindrops, snowflakes start to fall when they are too big to stay up in the cloud. The ice crystals fall at different speeds, because of their varying shapes. As they fall they collide and clump together to form snowflakes. This is called aggregation.

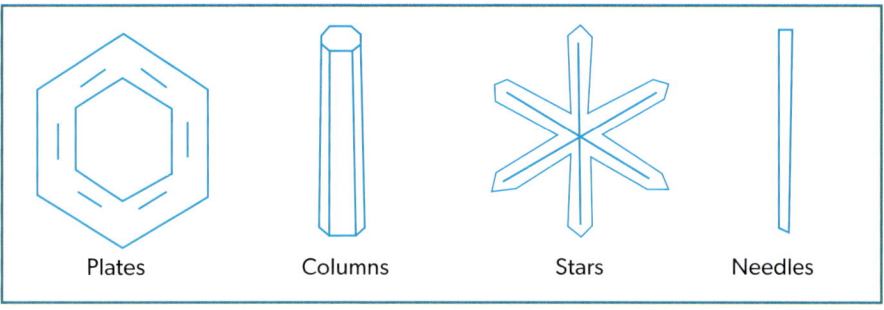

Most snowflakes are regular six-sided shapes. If they spin as they fall they keep their regular shape, but if they fall lopsidedly they can become an irregular shape.

Smaller snowflakes come from ice crystals and super-cooled water droplets bumping into each other and freezing together. If that happens often enough they lose their shape and form hail.

MetService issues a heavy snow warning if snow is expected to settle below 1000 metres in the North Island, South Canterbury and Otago, or below 500 metres in the rest of the South Island. New Zealand's maritime climate means that snow here is wetter, denser and heavier than snow in a drier climate, like Canada's. This can make it more damaging to power lines and infrastructure like buildings and more likely to disrupt air, road and rail travel.

HOW RAIN FALLS

- Inside clouds ultra-fine water droplets stay afloat on invisible, upward-moving air currents.
- As they move around the cloud they collide with each other and combine to form bigger droplets. This is called collision and coalescence.
- When the drops grow too big and heavy to stay up, the water droplets start to fall. As they fall, larger droplets can draw smaller ones into their wake, merging with them. The bigger the raindrop, the faster it falls.
- If the temperature is low enough in high parts of the cloud, super-cooled water droplets and ice crystals form.
- Ice crystals can form in different ways: directly from water vapour (sublimation); by colliding with water droplets which freeze onto them (riming) – this is how hailstones can begin to form; or by crashing into other ice crystals and sticking together (aggregation). High clouds often contain both super-cooled water droplets and ice crystals.
- Super-cooled water droplets are droplets of liquid water at below freezing temperature. Scientists think they form in the absence of condensation nuclei. Once they come into contact with ice crystals they freeze.

High cirrostratus over Lake Pukaki looks more like a milky veil over the sky than a cloud. The ice crystals reflect sunlight and scatter it, like tiny shimmering diamonds. Cirrostratus often means a change in the weather is on its way.

MAN-MADE SNOW

Ski fields use snow machines to make extra snow but the air has to be freezing cold and just the right level of humidity. Snow makers spray a mist of water droplets into the air, the droplets freeze into ice crystals, then they clump together and form snowflakes as they fall to the ground.

JET STREAM WINDS

Changes in the usual flow of jet streams can make extreme weather events worse. The polar jet streams usually circulate above the poles, holding the cold polar air, the polar vortex, in place. An unusually strong Antarctic polar vortex gave Aotearoa the warmest winter on record in 2021.

The strength of a jet stream depends on temperature contrasts in the upper atmosphere. When the contrast lessens, jet streams weaken and their track starts to weave and buckle. As the jet stream winds meander from their usual paths these bulges draw in cold air from further south or warm air from further north, in our hemisphere. This brings unusual weather, which can be abnormally hot or cold, wet or dry. When a weather system gets blocked and stays in place for a long period, we see dramatic weather events like heat waves, cold snaps, extreme flooding or drought, breaking more weather records. Climate change may affect the balance of temperatures in the upper atmosphere and make the jet streams more wavy. Waves in the jet stream can also cause storm systems or heat waves to move more slowly or get stuck in place. Climate change does not mean even heating of the planet — it can bring extreme cold snaps too. A blast of unusually cold air from Antarctic storms can reach New Zealand, as it did in the winter snow storms of 2011 when a polar outbreak brought the heaviest snow to Wellington in 30 years.

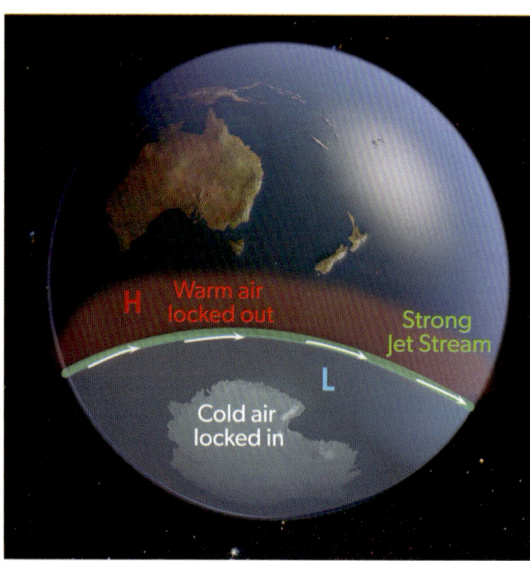

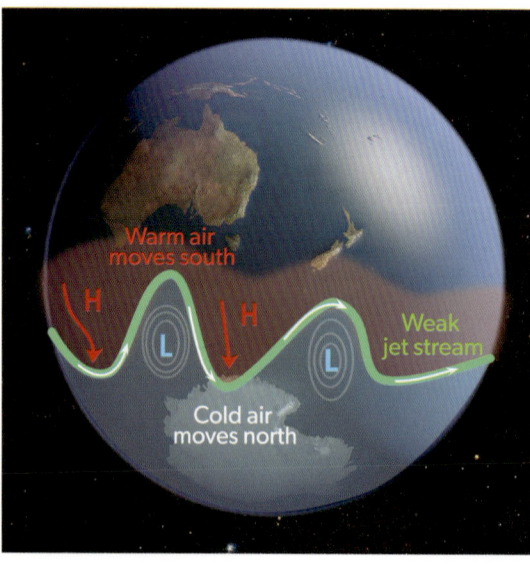

FROST

Frost can happen almost anywhere in Aotearoa, although the number of frost days is decreasing in most areas, especially at the top of both islands. A frost day is one with a daily minimum air temperature of below zero. Frosts are less common on the coast and most common in central inland areas. On the North Island's Central Plateau and in the Mackenzie Basin (Canterbury) and Central Otago, heavy frosts can damage young fruit, developing buds and blossom and cause poor harvests for growers.

The most severe frosts occur in the heart of the South Island, in Central Otago, with its continental climate and greater extremes of weather. Orchardists and wine-growers use frost pots, wind machines and even helicopters to protect their crops from frost by warming and stirring up the air.

> **MARAMATAKA** Maramataka, the traditional Māori lunar calendar, marks the passage of the seasons and the movements of the stars, and acts as a guide to fishing, planting and harvesting, according to the phases of the moon. Each month is linked to a different star or group of stars and each night of the month has its own name. Now that the times of year when fish migrate or pollination takes place are shifting, the best times for fishing and planting are changing too and maramataka is having to adapt to the changing climate. (TE PAPA)

Frosts occur when there is a High on the weather map, on cold, clear, windless nights when you can see the stars and the temperature falls below zero. With no cloud cover to trap warmth, the heat of the day radiates from the surface back into space. On cool, cloudy nights, when the air cools down to its dew point, water droplets condense onto the ground and nearby surfaces and form dew. Under cold and cloudless night skies, when the land and surfaces cool to below freezing, the water vapour in the air freezes onto surfaces as ice crystals. Sometimes the frost does not appear till around sunrise, when the heat loss is near its greatest. We get more frosts in winter and early spring when nights are longer and Earth cools down more.

Sometimes Central Otago experiences hoar frosts, which form when water vapour freezes onto surfaces instantly as ice crystals, without becoming water droplets first.

While frost can be damaging to crops, some fruit and nut trees need a period of winter cold to develop properly. Winter, as defined by the average daily temperature, is at least a month shorter in Aotearoa than it was eighty years ago. It tends to start later and finish earlier. This affects the life cycles of animals, plants and insects, which in turn affects pollination and harvest times.

> **WHAT IS A CONTINENTAL CLIMATE?**
> Regions that are surrounded by land rather than sea have a continental climate. The climate of Central Otago has continental characteristics: hot summers and cold winters; hot days followed by cold nights; a huge range of temperatures over 24 hours (sometimes varying by 20 degrees); a 'desert' region with very low rainfall. Central Otago often has barely 400 millimetres of rain a year. Alexandra had a record high of 38.7°C on 30 January 2018 while Ranfurly, 56 km away, has New Zealand's national record low of -25.6°C on 17 July 1903.

FOG AND MIST

Fog is really stratus cloud on the ground. Fog often forms in valleys and hollows in the land where cold air collects and there is plenty of moisture in the air, especially in the lowlands of the Waikato and the King Country. Taumarunui, where the Whanganui River joins the Ongarue River, has about 100 days of fog a year.

> **WHAT'S THE DIFFERENCE BETWEEN FOG AND MIST?**
>
> The difference lies in the amount of tiny water droplets in the air and how much light scatters from them and restricts how far you can see.
>
> **Fog** you cannot see more than 1000 metres ahead.
>
> **Mist** you can see more than 1000 metres ahead.

RADIATION FOG

On a clear night after a sunny winter's day, when Earth's heat is radiated out into space, the ground and the air directly above it cool. This cooling spreads slowly upwards through the lower layers of the atmosphere and if there is enough moisture in the air and the air is still, the water vapour it holds condenses into dew or frost, depending on the temperature. When there is a light

Widespread frost over Otago with fog in the valleys.

breeze, however, the air mixes and becomes cooled a little higher above the surface. The water droplets suspended (floating) in this air then condense to form fog. This is called radiation fog (the most common kind of fog in Aotearoa), because it happens when Earth's heat radiates out to space at night.

Radiation fog usually clears quickly after the Sun rises. The Sun's heat reaches through the fog and warms the Earth below. This, in turn, warms the air just above the surface, which makes the water droplets evaporate from the ground up and so the fog 'lifts'.

STEAM FOG

Steam fog forms a thin layer over water, when cold air moves over warmer water. On a clear night, the land loses heat much faster than

the water so the water remains warm in comparison. When the Sun comes up, the air just above the water warms enough for a small amount of moisture to evaporate. If it evaporates fast enough, the air quickly becomes saturated (unable to hold any more moisture) and the water droplets soon condense again, forming fog.

ADVECTION FOG

When advection fog (*advection* means air moving horizontally) forms above the sea it is known as sea fog. Advection fog often occurs at the coast, when warm, moist (subtropical) air from the north moves in over a cooler land or sea surface and is cooled from below. Once the air above the cooler surface cools, the moisture it holds condenses to form fog. Advection fog can last for much longer than radiation fog, even in quite windy conditions. It can form in clear or cloudy skies, by day or night and is the kind of fog that causes problems at Wellington Airport. The fog only clears when the wind direction changes, bringing in a drier air mass.

GLACIERS

A glacier is a slow-moving river of ice, made of snow that has fallen high in the mountains, year after year. Over many years the snow gets packed down and compressed into ice which, under its own weight, gradually flows down steep valleys. Glaciers depend on fresh snowfall to advance (grow). You can track this in the position in the valley of the terminus or lower end of the glacier, where the ice melts.

HOW FAST DO GLACIERS MOVE? NZ glaciers have been estimated to move at rates ranging from 20–600 metres per year.

Glaciers are indicators of climate change, both past and present, as their growth is affected by changes in air temperature and in precipitation in the mountains. Glaciers only grow if the amount of snowfall is greater than the amount of ice melting.

New Zealand has more than 3000 glaciers, mostly in the Southern Alps, but at least 18 have been charted around Mount Ruapehu.

A glacier near Mount Aspiring in the Southern Alps.

> **RATE OF CHANGE** In 2019, Stats NZ recorded that as 'at June 2019, the volume of ice in New Zealand was 35% lower than the estimate for the year ended June 1995'. From '1997 to 2016, 15.5 km³ of ice was lost, enough to fill Wellington Harbour 12 times'.

Since 1950 almost all of the world's glaciers have been shrinking, faster than at any time in the last 2000 years. Their melting has almost doubled in speed over the past 20 years and accounts for more than one-fifth of sea level rise. In spring, meltwater from snow flows into rivers, which provide water for irrigation and feed the lakes that power our hydro-electric dams.

Every year since 1977 scientists from NIWA and Victoria University of Wellington chart the position of 50 New Zealand glaciers in their 'End Of Summer Snowline Survey'. They take aerial photos, including infrared images, and make 3D models to assess the extent of the glaciers and the thickness of the ice. Rises in the surface temperature of the Tasman Sea contribute to rising air temperature and serve to speed the melting. In a feedback loop, airborne pollution like soot and dirt from Australian bush fires drifts across the Tasman and settles on the Alps; the darker surface of the ice and snow absorbs rather than reflects the Sun's radiation and this, again, further speeds the melting. The Brewster Glacier lost 13 billion litres of water in the eight years to 2021. This is nearly enough water for every New Zealander to drink a litre a day over that timeframe.

Lakes Pūkaki and Tekapo get their milky turquoise colour from fine sediment in glacier meltwater. Many South Island lakes are long and narrow because they were carved out by glaciers.

THE CLIMATE CRISIS

Even taking predicted climate changes into account, in the past decade New Zealand has experienced more extreme weather than expected. The number of warm days (days where the maximum temperature is above 25°C) has increased at 24 of NIWA's 30 measuring stations around the country.

WHAT CAN YOU DO?
Te ao Māori — the Māori perspective
We are part of the natural world, the environment, not separate from it. Planet Earth is a taonga, a treasure, for us to respect and conserve. We can all act as kaitiaki, guardians, and protect the mauri, the life force of the planet, to keep it strong and healthy. We need to be mindful of our impact on sky, land and sea and think about what we choose to eat, buy and do.

> One Māori proverb says, 'Tūngia te ururoa kia tupu whakaritorito te tutū o te harakeke', and means, 'set the overgrown bush alight, and the new flax shoots will spring up': in order to change we may need to leave some ways behind so we can do things differently.

ACTIVISM
Activism is taking direct action to support an issue or bring about change in the way things are run. You may not be old enough to vote yet but it is your future and you need a world you can live in. Things you can do now:

- Join with friends to research local issues — just because things have always been done a particular way, it doesn't mean it always has to be like that. We need to adapt to tackle the climate crisis.
- Sign petitions (e.g. with ideas for local transport solutions).
- Write emails or letters to your local council, your local MP or to the Prime Minister.
- Check out the Climate Commission's recommendations.

CONSUME LESS, USE LESS, WASTE LESS, BUY LESS
At home and in family life you can:
- Use less electricity — turn off lights and unplug electrical items when you are not using them.
- Read books, play cards and board and outdoor games.
- Don't waste food. Compost scraps. Food production emits lots of greenhouse gases — about 30% of your carbon footprint can come from food you eat. Sign up to New Zealand websites that match you to people who want food scraps (www.sharewaste.org.nz)
- Avoid sending anything to landfill. Landfills give off greenhouse gases like CO_2 and methane. Ask yourself: can I donate my item? Can I fix this? See if there's a repair cafe near you.
- Before travelling, ask yourself: Do I need to travel? Can I bike, walk, bus, go by train or ride-share?
- Fashion accounts for more emissions globally than flying and shipping combined. Ask yourself: do I really need this? Think about the item's carbon footprint. Buy from op shops and local markets; arrange clothes swaps with friends. Learn to mend, knit and sew.
- Buy local and choose fruit and vegetables that are in season. Lots of small choices add up: less food travelling round the globe and the country means lower emissions.
- Eat foods with lower carbon footprints: more plant foods (unprocessed) and less meat and dairy.

WHAT GOVERNMENTS CAN DO TO REDUCE EMISSIONS
- Regulate against wasteful packaging.
- Regulate to make manufacturers recycle products and to stop them making it impossible to repair products.
- Increase public transport, especially rail networks, and cut roading projects.
- Encourage electric vehicle uptake, as well as car sharing and leasing arrangements.
- Increase incentives and funding for green energy and transport and green housing development and promote regenerative farming.
- Stop issuing permits for fossil fuel exploration.
- Stop investment in fossil fuels.

START SMALL BUT START NOW

Remember the Greta Thunberg effect — a worldwide movement started with one brave individual's solo protest. To keep planet Earth habitable we need to rethink our priorities. Every action, every decision, has to take the planet into account.

START SMALL, THEN THINK BIGGER

- Be brave — don't be afraid to speak out and share ideas. Start with your family and friends, then your school or sports group or club. Try new ideas such as eating meat-free one day a week.
- Plant trees to absorb CO_2. Young trees absorb more carbon than old ones.
- Gather seeds from native trees and grow them yourself.
- Ask your school if you can plant a veggie garden.
- Hold a bake sale to fundraise for a school or club compost bin and fruit trees to plant.
- Avoid 'doomscrolling' — dwelling on negative news. People are going to have to change the way they live, like it or not.

THINK BIGGER

If we all reduce individual carbon footprints and encourage others to do the same, our actions add up, but they aren't enough on their own. The world needs large scale changes to keep global heating within liveable limits. Governments are usually afraid of making changes they fear will be unpopular. So, as well as making changes in our everyday lives, we need to tell our local and national government representatives that we support them making the far-reaching urgent changes our world needs — in transport, in food and energy production, in manufacturing, in our buildings and homes — in every aspect of our lives.

In 2020 the Covid-19 pandemic showed that governments around the world CAN act fast in a global emergency. Countries locked down and there was a 5.4% reduction in global emissions that year. (But they rose again once countries began opening up.) At the end of 2021, the world needed to achieve those reductions every year for the next nine years to reach the target of limiting global temperature rise to 1.5°. We need to cut global emissions by about half by 2030. At the start of 2022, New Zealand was planning to buy two thirds of its carbon credits from other countries to meet its emissions targets, rather than reducing our emissions.

INDEX

Aerosols 30, 60, 63, 71, 86
Air pressure 9, 12, 46, 49, 52–56, 57–58, 68–69, 72, 78
Albedo 22, 30
Anthropocene 25
Anticyclone 69–73
Atmosphere 8–10, 13–19, 21–22, 26–27, 40–44, 83, 90
Aurora 39

Biosphere 19, 26

Carbon cycle 11, 18, 27, 44
Carbon dioxide 21–27, 31–33, 42–44, 98, 99
Carbon footprint 32, 98, 99
Carbon neutral 31
Carbon sink 42, 43, 44
CFCs, chlorofluorocarbons 31, 35
Climate change 24, 25–29, 83, 84, 92, 95, 97–99
Clouds 21, 42, 59–60, 71, 73, 76, 81, 86, 89, 92
Convection currents 13, 16, 41, 42, 57
Coriolis effect 14, 51, 69, 70
Corona, CMEs, coronal mass ejections 36, 37, 38
Cryosphere 19, 26
Cumulonimbus cloud 59–60, 61–63, 65, 66, 76

Depression 48, 73–75, 83
Dewpoint 62, 85, 92
Drought 26, 41, 42, 83, 90

Earth system 15, 30
El Niño 51, 52–56
Electromagnetic spectrum 37

Feedbacks 26–27, 28, 29
Ferrel cell 16–17, 69
Flooding 26, 42, 48, 51, 83, 84, 90
Fog 93–94
Forcings 28, 30
Forecasting 23, 67, 80
Fossil records 25–26
Fronts 62, 72, 74–77, 83
Frost 56, 91–92, 93

Geomagnetic storms 38, 39

Glaciers 14, 19, 27, 29, 95–96
Glasgow Climate Pact 32, 44
Global heating 33, 36, 41, 42, 51, 81, 83
Greenhouse gases and effect 21, 27, 30, 43, 98

Hadley cell 16, 17, 69
Hail 63, 65, 74
Heat waves 56, 71, 84, 90
High *see* anticyclone
Hydrosphere 19, 26

Ice ages, interglacials 25, 26–27
Ice, ice sheets 22, 30, 47
Inversion 10, 70
Ions, ionosphere 11, 37, 39
IPCC 30, 31, 32–33, 45, 47, 99
Isobars 72, 74, 75
Isothermal layer 11

Jet stream 16, 28, 62, 69, 74, 89–90

Kyoto Protocol 36

La Niña 51, 52–56
Latitude 5, 6–7, 14, 29
Lightning 63–65
Lithosphere 19, 26, 44
Low *see* depression

Magnetosphere 11, 38
Marine heatwave 47
Maritime climate 6, 8, 68–69, 88
Methane 21, 31, 32, 98
MetService 23–24, 80, 84, 88
Microburst 62
Microclimate 5
Milankovitch cycles 29
Models (climate) 28, 30
Montreal Protocol 36
Mountains 5, 79, 80–82, 83, 85

Net zero 31, 33
NIWA 24, 30, 36, 37, 39, 84, 97

Ocean acidification 43
Ocean deoxygenation 46
Oceans 8, 11, 13, 29, 40–47

Orographic lifting 85
Ozone, ozone hole 11, 34–37

Palaeoclimatology 26–27
Polar vortex 35, 89
Pressure gradient 57, 78

Rain, rainfall 5, 13, 41, 42, 49, 52–55, 62–63, 74–77, 83–88
Ridge 72–73, 75

Sea breeze 57, 58, 62
Sea level 25, 26–27, 45, 47, 96
Snow 63, 84, 87–90
Solar flares, storms, wind 38
Solar radiation (sunlight) 21–22, 27, 28, 29, 34, 37
Southeast trade winds 8, 13, 48, 52–55
Speed of change 14, 25, 26, 28, 29, 41, 43, 96
Storm surges 45, 46
Stratosphere 9–11
Sun, sunshine 10, 13, 29, 36
Sunspots 28

Temperature, temperature gradient 9, 10–12, 21–22, 25, 26, 33, 56
Thermals 81
Thunderstorms 61–67, 74, 76
Tipping point/cascade 28
Tornadoes 63, 66–67, 84
Tropical cyclones 48–51
Tropics, tropical climate 6, 7, 68–69
Troposphere 9–12, 69, 79
Trough 74, 75

UV (ultraviolet) 9, 34–37

Volcanic eruptions 25, 26, 27, 29, 30, 44, 60

Walker circulation 52–55
Water cycle 11, 18, 27
Water vapour 11, 21, 62, 70, 71, 73, 83, 86, 87, 88, 92
Westerlies 8, 13, 14, 36, 48, 78–79, 81, 85–86
Wildfires 27, 31, 63, 84
Wind 8, 51, 52–55, 66–67, 74–77, 78–82, 84